INTRODUCTORY TO RAY OPTICS

INTRODUCTORY TO RAY OPTICS

Arpit Saxena

CYBER TECH PUBLICATIONS
4264/3, Ansari Road, Daryaganj, New Delhi-110002 (India)
Ph.: 011-23244078, 011-43559448 Fax: 011-23280028
E-Mail: cyberpublicationsdelhi@yahoo.com
Website: www.cybertechpublications.com

INTRODUCTORY TO RAY OPTICS

Arpit Saxena

First Edition 2010

Asian Offset Printers

Published by :

G.S. Rawat for **Cyber Tech Publications**
4264/3, Ansari Road, Daryaganj, New Delhi-110002 (India)
Ph.: 011-23244078, 011-43559448 Fax: 011-23280028
E-Mail: cyberpublicationsdelhi@yahoo.com
Website: www.cybertechpublications.com

PREFACE

The subject matter of this book has been discussed in such a simple way that the students will find no difficulty to understand it. This books has been specially written to the requirements syllabus of the students of all Indian Universities. Chapter of this book contains complete theory and fairly large number of solved examples.

The proofs of various theorems and examples has been given with minute details.Sufficient problems has also been selected from various universities examination papers. and the author shall be thankful to the readers who point out errors and omissions which inspite of all care might have been there.

The author will feel amply rewarded if the book serve the purpose for which it is meant. The student suggestion some errors for the improvement of the book are always welcome.

—Author

Contents

1

CONJUGATE POINTS OF THE LENS COMBINATION

INTRODUCTION

We study to two typical applications, *telephoto lens* and telescopic combinations and take up some other combinations, eyepieces, at a later stage. Optical instruments employ a variety of lens combinations to achieve the desired functions. We cannot study all the different combinations here.

TELESCOPE LENS

A simple astronomical telescopic system consists of a combination of two single convex lenses. The back focal plane of first lens is made to coincide with the front focal plane of the second lens. Therefore, the separation of the lenses d is given by $f_1 + f_2$. Consequently,

$$\Delta = f_1 + f_2 - d = 0,$$

and the equivalent focal length of the combination,

$$f = \frac{f_1 f_2}{\Delta} = \infty \text{ . Further, } \alpha = \frac{fd}{f_2} = \infty \text{ and } \beta$$

$= -fd/f_1 = \infty$. Thus, all the cardinal planes (focal and principal planes) are at infinity. It means that a telescope gives images at infinity for distant objects; this is the condition appropriate for viewing by a relaxed eye.

Telescopes perform two functions. One is to produce an enlarged image of a distant object, and the other is to gather more light from the object than is possible with the unaided eye.

Thick Lens in Air

In Fig. 6.1, OB and O'C represent two spherical surfaces of radii of curvature R_1, R_2, separated by distance POO' = (and enclosing a medium of refractive index n. This is called a 'thick lens' if t is not *negligible in proportion* to R_1, R_2.

To obtain the focal length and positions of cardinal points, consider an incident ray AB parallel to the principal axis. In two refractions it bends as BC, CF' meeting the axis at F', which is the *focal point of image space.* If we produce AB and CF' they meet at P. This is where the emergent ray has the same height h as the incident ray. Hence, PH' gives us the *principal plane of image space.* H'F' = f' is the *focal length of image space.*

PLANES AND DISTANCES—CONJUGATE POINTS

In Fig. 1 all the rays coming from an *object point* B are brought to a focus at B'. Similarly, the rays from point A are brought to a focus at A'. A' B' is the *image* of the linear object AB. It follows from the principle of reversibility of light rays that the object and image positions are interchangeable. Thus, when the image A' B' becomes the object and AB becomes its image. The *object and image are therefore conjugate.* Any pair of object and image points A and A', as in Fig. 1, are *conjugate points.* Planes on which the conjugate points lie are *conjugate planes* and the distances from the vertex to these planes are *conjugate distances.*

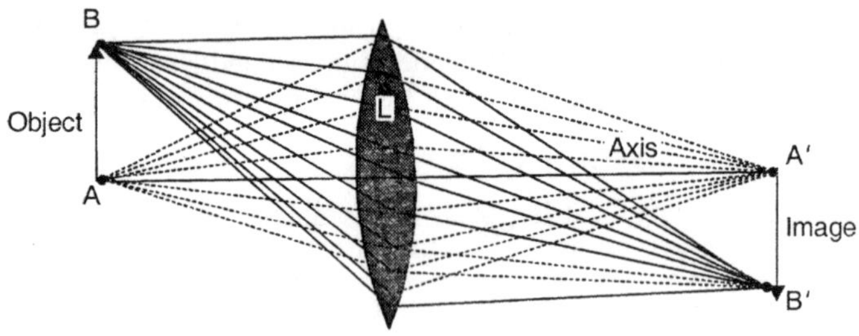

Fig. 1

TELEPHOTO LENS

To photograph a distant object, a camera should have a convex lens of long focal length. The image size for a distant object is directly proportional to the focal length of the lens. Therefore, to produce a large image on the film surface, the distance between the lens and photographic film should be large. Thus, a magnified image requires a long camera. On the other hand, the overall size of the camera should be reasonable and manageable. We can achieve this *only if* the image plane is close to the lens. These two contradictory requirements, of long focal length and short lens-to-focal-plane distance, can be resolved if we use a lens combination in stead of a single lens. Such a lens combination used in photographing distant objects is known as a *telephoto lens*. A *telephoto lens* consists of a convex lens of focal length f_1 and a concave lens of focal length f_2 separated by a distance slightly greater than $f_1 + f_2$ The concave lens L_2 is placed in the original position of the lens in the camera whereas the convex lens L_1 is in front of L_2, as shown in Fig. 6.8. The combination of these two lenses acts as single convex lens of large focal length. The equivalent focal length of the combination is given by

$$\frac{1}{f} = \frac{1}{f_1} + \frac{1}{f_2} - \frac{d}{f_1 f_2}.$$

The focal length f of the system is measured from the second principal plane H_2P_2 and the lens to focal plane distance, L_2F_2 is measured from L_2 to the photographic film. The distance from the first lens to the photographic film, L_1F_2, determines the overall length of the camera. What we require is shorter L_1F_2 and longer P_2F_2. The position of the second principal plane is determined by the intersection of a ray, AB, parallel to the axis in the object space with its conjugate ray, CF_2, in the final image space. It is seen that the second principal plane H_2P_2 lies well in front of the first lens L_2. Therefore, the equivalent focal length of the system, $f = P_2F_2$, is quite large. The lens-to- focal-plane distance is reduced because the back focal length, $f_b = L_2F_2$ is shorter. Thus, the overall length of the photo lens plus camera is made shorter.

To fully appreciate the function of the lens combination in the telephoto lens, let us consider the following example. A telephoto lens consisting of a convex lens of focal length 10 cm and a concave lens of focal length (–3 cm), separated by a distance 8 cm.

The focal length of the combination,

$$f = \frac{f_1 f_2}{f_1 + f_2 - d} = \frac{10\text{cm} \times (-3\text{cm})}{10\text{cm} - 3\text{cm} - 8\text{cm}} = +\,30\text{cm}$$

$$\therefore \quad f = P_2F_2 = 30 \text{ cm}$$

$$\beta = -\frac{fd}{f_1} = -\frac{30\text{cm} \times 8\text{ cm}}{10\text{cm}} = -\,24\text{ cm}$$

$\therefore \quad L_2P_2 = -\beta = 24$ cm

Lens-to-focal-plane distance

$$L_2F_2 = P_2F_2\ P- P_2L_2 = (30 - 24)\text{ cm} = 6\text{cm}.$$

$$\alpha = \frac{fd}{f_2} = -\frac{30\text{cm} \times 8\text{ cm}}{-3\text{cm}} = -80\text{cm}$$

$\therefore \quad L_1P_1 = -80$cm.

Overall length of the camera $L_1F_2 = d + L_2F_2 = (8 + 6)$ cm = 14 cm.

If a single convex lens is used the length of the camera would have been 30 cm. The lens combination reduced the camera length to 14 cm, *i.e.* approximately by half.

IMAGE TRACING

We may use graphical ray tracing to determine the position of the image formed by a lens. To find the image, we take the help of characteristic rays shown in Fig. 2.

1. One is the ray parallel to the principal axis, which after refraction, passes through focal point F_2.

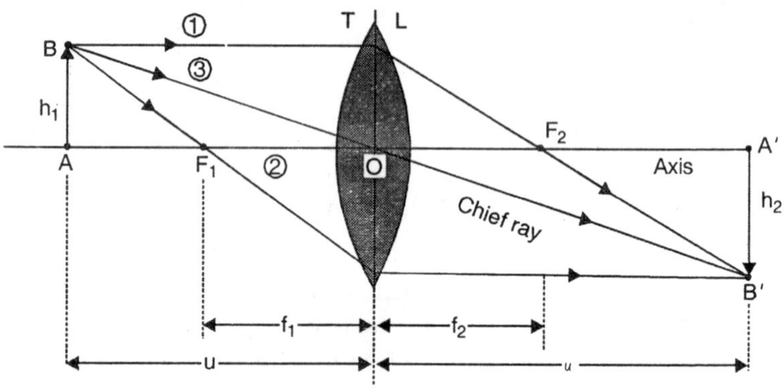

Fig. 2

2. Second ray is the ray that passes through the first focal point F_1 of the lens; after refraction, it travels parallel to the principal axis.

3. The third ray, usually called *chief ray* goes through the optical centre of the lens and emerges without deviation.

SOME RULES FOR OBTAINING IMAGES FORMED BY CONVEX LENSES

When an object is placed in front of a convex lens, an image is formed. The image is formed at that point where at least two refracted rays meet or appear to meet. The following two light rays are usually used to locate the images formed by convex lenses :

Rule 1: An incident ray parallel to the principal axis of a convex lens, passes through the focus of the lens after refraction.

Rule 2: An incident ray passing through the optical centre of a convex lens goes straight after refraction through the lens. It does not get deviated. Another ray of light which is also sometimes used to locate the images formed by convex lenses is the following :

Rule 3: An 'incident ray passing through the focus of a convex lens, becomes parallel to the principal axis after refraction through the lens. This is just the opposite of rule 1 given above.

Thus, we have three characteristic rays whose paths are known. Using any two of the three characteristic rays, we can readily determine the image of any object-point or of any extended object. Suppose a ray is incident on a lens at an arbitrary angle, as shown in Fig. 4.4. To find where it forms the image, we draw a secondary optical axis parallel to the ray and a plane perpendicular to the principal axis and passing through the second focal point of the lens; this plane is the second focal plane. It is readily seen that just as a paraxial beam of light parallel to the principal optical axis converges at the focal point F_2 so the beam of light parallel to the secondary optical axis converges at a point on the focal plane. Thus, light rays incident on the surface of a convex lens at a particular angle to its *axis*, are redirected towards a particular position in the *focal plane* of the lens. We can therefore regard a convex lens as an *angle-to-position converter* for light.

Location of the Image

A convex lens produces a real or virtual image depending on the location of the object. A concave lens *always* produces virtual images of real objects. The behaviour of a convex lens pictorially. As the object approaches the lens, it is seen that the real image moves away from it.

(i) When the object is very far away, the image is just to the right of the focal plane. The image is real, inverted, and smaller in size than the object ($m < 1$).

(ii) As the object approaches the lens, the image moves away to the right of the focal plane getting larger and larger. The image is real and inverted. This is the configuration for cameras and eyeballs.

(iii) When the object is at 2F, the image is real, inverted and of the same size as the object (m = 1). This is the configuration of a photocopier.

(iv) When the object is in between 2F and F, me image is enlarged (m > 1), real and inverted. This configuration corresponds to the film projector.

(v) When the object is precisely at F, there is no image as the emerging rays are parallel and in effect the image is at infinity.

(vi) With the object closer in than one F, the image reappears. It is virtual, erect and enlarged (m > 1). This is the configuration of the magnifying glass.

It may be noted that as an image-forming device, a convex lens is similar to a convex mirror. On the other hand, a concave lens resembles a convex mirror.

Thick Lens

A thick lens is a physically large lens having two spherical surfaces separated by a distance, which is not negligible in comparison to the radii of curvature of the spherical surfaces. In other words, a thick lens is a lens whose thickness cannot be treated as small in comparison to its focal length. We can determine the position and size of the image of an object by considering refraction at each surface successively, but the procedure is comparatively tedious. The position and size of the image can be determined *directly* if we know the cardinal points of the thick lens.

PRINCIPAL POINTS A THICK LENS

The thick lens, AB, bounded by two spherical surfaces separated by a distance. Let a ray OA travelling parallel to the principal axis be incident on it. After refraction through the lens, the ray OA passes through the focal point F, in the *image space* of the lens. The actual ray is refracted at each surface of the lens and follows the path $OABF_1$. If we extend the ray OA forward and the ray BP^ backward, they meet each other within the lens at H_2 Now, we can describe the refraction of the incident ray OA in terms of a *single refraction* at a plane passing through

H_2. A plane drawn through the point H_2 and perpendicular to the axis may be regarded as the surface at which refraction takes place. Such a plane H_2P_2 is called a *principal plane.* The point P_2 is called the *Second Principal Point.* Thus, the two consecutive deviations of the light rays caused by the two surfaces at A and B of the thick lens are equivalent to a single refraction at H_2, taking place at the *principal plane.* H_2 is the principal plane in the image space and is called the *second principal plane.*

We can locate the first principal plane H_1P_1 and first principal point P_1 in the *object space.* Consider the ray TQ travelling parallel to t_1 principal axis and incident on the second refracting surface at point Q. The ray after refraction through the lens along TS, emerges from the refracting surface through point S along SF_1. The F_1S and TQ when produced intersect at H_1. A plane passing through H_1 and perpendicular to principal axis is called the *first principal plane.* The point of intersection, P_1 of the first principal plane, H_1P_1 with the principal axis is called the *first principal point.*

The two incident rays AB and F_1S are directed towards H_1 and after refraction seem to come from H_2. Therefore, H_2 is the image of H_1. Thus, H_1 and H_2 are the *conjugate points* and the planes H_1P_1 and H_2P_1 are a pair of *conjugate planes.* It is also seen that

$$H_1P_2 = H_1P_1$$

Hence, the lateral magnification of the planes is +1 .Thus, the first and second principal planes are planes of unit magnification and are therefore called *unit planes* and the points P_1 and P_2 are called *unit points.*

Note : The principal planes are conceptual planes and do not have physical existence within the lens.

FOCAL POINTS

The *first focal point* is a point on the principal axis of the lens such that a beam of light passing through it is rendered parallel to the principal axis after refraction through the lens. The *second focal point* is a point on the principal axis of the lens such that a beam of light travelling parallel to the principal axis of the lens, after refraction through the lens, passes through it.

The planes passing through the principal focal points F_1 and F_2 and perpendicular to the axis are called *first focal plane* and *second focal plane* respectively. The main property of the focal planes is that the rays starting from a point in the focal plane in the object space correspond to a set of conjugate parallel rays in the image space. Similarly, a set of parallel rays in the object space corresponds to a set of rays intersecting at a point in the focal plane in the image space.

The distance of the first focal point from the first principal point, *i.e.*, $F_1 P_1$ is called the *first focal length*, f_1, of the lens and the distance of the second focal point from the second principal point, F_2P_2, is called the second focal length, f_1, f_1 and f_2 are also known as the focal lengths in object space and image space respectively.

When the medium is same on the two sides of the lens $f_1 = f_2$ (numerically).

EXTENDED OBJECT AT INFINITY

Let AB be a ray parallel to the axis and coming from infinity. After refracting along BG, it emerges at G at the second surface and goes along GF_2. Let H_1P_1 and H_2P_2 be the first and the second principal planes. Let R_1 be the radius of curvature of the first surface and R_2 be the radius of curvature of the second surface. Let μ be the refractive index of the lens material and let t be the thickness of the lens. Let f_1 and f_2 be the first and second focal lengths of the lens respectively and let the medium on either side of the lens be air.

NODAL POINTS

Nodal points are points on the principal axis of the lens where light rays, *without refraction*, intersect the optic axis. The planes passing through the nodal points and perpendicular to the principal axis are called the *nodal planes*. Whereas the principal planes are planes where all refraction are assumed to occur, the nodal planes are planes where refraction does not take place.A thick lens with the help of its cardinal planes. It is seen from the Fig. 6.4 that a ray of light directed towards one of the nodal points, N_1, after refraction through the lens emerges out from the second nodal point, N_2, in a direction parallel to the incident ray. The distances of the nodal points are measured from the focal points.

The formation of the image of an object OO'. An object OO' is placed on the axis of a thick lens. The paraxial incident ray O' A travels parallel to the axis and is incident on the lens at A. The ray after refraction through the lens passes through the second focal point F_2. Another paraxial ray

O' F_1Q starting from O' passes through the first focal point F_1 and after refraction through the lens emerges at E parallel to the principal axis. Another paraxial ray O' N_1 starting from O' passes through the first nodal point N_1 and emerges at along N_2D after refraction through the lens. The three emergent rays intersect at I' which is the image of O'. The line I I' drawn perpendicular on the axis represents the image of the object OO'.

In *symmetrical lenses*, the principal planes are *symmetrically* located and lie within the lens. In a plano-convex and even more so in a meniscus, the two principal planes move to one side and may even move outside of the lens, towards the side of steeper curvature.

NODAL SLIDE

A nodal slide is a particular type of horizontal metal support for a lens system, which is capable of rotation about a vertical axis. The nodal slide provides a convenient method for locating the focal and nodal points and for determining the focal length of a lens system.

Principle: Any optical system has two nodal points N_1 and N_1. An incident light ray directed towards N_1, after refraction through the system, proceeds from N_2 in a direction parallel to the incident ray. The method of locating the nodal points with the help of the nodal slide involves the following principle.

"If a parallel beam of light is incident on a convergent lens system, it forms an image on a screen held at its second focal plane. When the lens system is rotated through a small angle about a vertical axis through its second nodal point, the image does not shift laterally and remains stationary."

Let us suppose a beam of parallel rays is incident on a coaxial lens system. The beam passes through the system and converges to the second focus F_2 and a real image is formed on the screen. Now let the system be rotated about a perpendicular axis through O, which lies between N_2 and F_2 Due to this rotation, the nodal points N_1 and N_2 shift to the positions N_1' and N_2' respectively.

A ray incident at N_1' travels along $N_1'I_1$ parallel to the incident ray. Since the incident beam is parallel, the image lies on the second focal plane. The point of intersection of the ray $N_1'I_1$ with the focal plane gives the new position of the image. Thus, when the axis of rotation lies between N_2 and f_2, a slight rotation of the system changes the position of the image.

Next let us say the system is rotated through a small angle about an axis passing through N_2. Then N_1 shifts to N_1' while N_2 remains fixed. A parallel ray incident at N_1' on passing through N_2 follows the path N_2I. Therefore, the image remains stationary, as seen from Fig. 5.12.

If the axis of rotation lies before N_1, as in Fig. 5.13, any small amount of rotation displaces N_1 and N_2 and consequently the image position changes. Thus, the position of the axis of rotation for which there is no displacement of the image can be found. It gives the second nodal point. As the media on both sides of the system being the same, the second nodal point coincides with second principal point.

DETERMINATION OF NODAL POINTS

The experimental arrangement consists of an optical bench on which four uprights are kept. They carry a plane mirror, the nodal slide, and a screen provided with a slit fitted with cross wires and lamp housing. Light from the lamp passes through the slit and is incident on the lens system. It is rendered parallel and on passing through the lens system it is reflected back by the vertical plane mirror. The reflected light once again passes through the lens system and is brought to a focus in the plane of the stilt The distance between the optical system and the screen is adjusted in such a way that a well-defined image of the slit is obtained on the screen. The image of the slit is formed slightly to one side of the slit itself. It is obvious that the centre of the slit is at the first focal point of the lens. The nodal slide carrying the lens system is now rotated through a small angle and it will be found that the image shifts side ward to the right or to the left. The nodal slide and its stand are then adjusted such that the direction of rotation of the image changes its sign and finally the image remains stationary for a slight rotation of the carriage. When this condition is reached, the axis of rotation passes through the second nodal point N_2. The other focal point nodal point can be determined by turning the nodal slide through 180° and repeating the experiment. Since the medium on both sides of the lens system is the same (air), the nodal points are also the principal points. The distance between the screen and the axis of rotation for the stationary image is an accurate measure of the first focal length of the lens system.

EQUATION OF THE LENS

We determine the cardinal points of a thick lens by assuming first that the object at infinity and then for the case of object located on the principal axis at a certain distance from the system. We shall find that the computations would yield identical results in both the cases.

SEPARATED LENSES AND MOVE OUT OF TWO-LENS SYSTEM

It is interesting to note how the positions of principal planes change when two positive lenses of equal power are first in contact and then are gradually separated.

When the lenses are in contact the two principal planes are close together within the system. As the lenses are separated, the principal planes also separate. When the distance between the lenses is equal to their focal length, the principal planes coincide with the lenses. When the distance is larger than the focal length, the planes have moved outside the lenses. And when the distance is equal to of the planes have moved out to infinity.

KINDS OF LENSES

There are two types—*convex* lens and *convave* lens. A convex lens is thicker at the center than at the edges a concave lens is thinner at the center than at the edges. A convex lens is a converging lens since a parallel beam of light, after refraction, converges to a point, F, A concave lens is called a diverging lens since ray coming parallel to the principal axis, after refraction, diverse out and seem to come from a point F. Within these two categories there is a variety of simple lenses.

NOTES

- For every lens, there is a point on the principal axis for which the rays passing through it are not deviated by the lens. Any ray passing through it emerges in a direction parallel to the incident ray. Such a point is called the *optical centre*. When the lens is thin and the radii of curvature of the two refracting surfaces are equal, then the geometric center of the lens becomes the *optical center* of the lens.

- The plane perpendicular to the principal axis of lens and passing through its focal point is known as the *focal plane*.

- The distance between the focal point F and the optical center of the lens is called the *focal length* of the lens.

- The length of the radius of curvature of surface is called the *radius of curvature, R.*

- The reciprocal of the length of the radius of curvature is known as the *curvature* C (C = 1/R). A lens has two centers of curvature and two radii of curvature, one for each refracting surface.
- The line joining the centers of curvature of the two curved surfaces is called the principal axis or simply axis of the lens.
- The points where the *principal axis* intersects the two refracting surfaces are called the front vertex and the back vertex.
- The point F to which a set of rays parallel to the principal axis is caused to converge (in case of convex lens) or appear to diverge (in case of concave lens) is the *principal focus*.
- When a point object or a linear object is placed on one side of a convex lens beyond the focal plane, an image is formed on the opposite side, as shown in Fig. 4.2. The distance from the front vertex to the object is called the *object distance u* and the distance from the back vertex to the image is the *image distance, v.*
- The *power* of a lens is the reciprocal of its focal length.

CONVENTION OF THE SIGN

The following convention of the signs is adopted for obtaining the relation between these quantities.

(a) The diagrams are drawn showing the incident light travelling from left to right.

(b) The distances are measured by taking the optical centre of the lens as the origin.

(c) The distances measured in the direction of the incident light are considered positive while those measured in the direction opposite to the incident light are taken as negative. (All quantities measured to the right of P are positive and all those to its left are negative.)

(d) Heights measured upward and perpendicular to the principal axis are taken as positive while those measured downward are considered negative.

(e) The angle made by a ray with the principal axis is taken to be positive if the ray has to be rotated anti-clockwise to become coincident with the axis, otherwise it is negative.

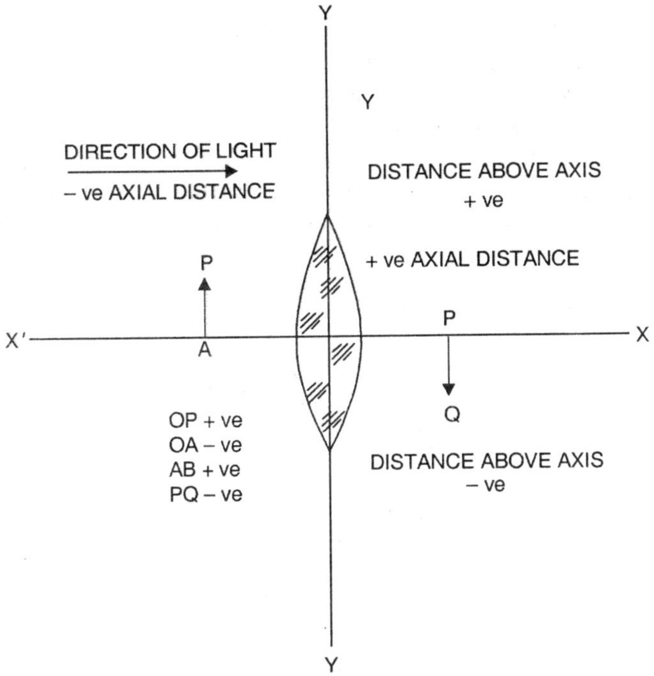

Fig. 3

THIN LENS

Lenses are broadly classified into *thin* and *thick* lenses. *A lens is said to be thin if the thickness of the lens can be neglected when compared to the lengths of the radii of curvature of its two refracting surfaces, and to the distances of the objects and images from it.* No lens is actually a thin lens. Yet many simple lenses commonly used can be treated as equivalent to a thin lens.

LENS EQUATION

A lens forms an image by refraction of light at its two bounding surfaces. Each surface acts as an image-forming component, and contributes to the final image formed by the lens. If we know the focal length of a lens and the position of an object, the position of the image can be determined either by using graphical construction (ray diagram) or using mathematical relation. It is not always convenient to draw a ray diagram.

We study here how the position of the image is determined using mathematical equation.

Now let us consider a *thin* convex lens XY with optical center at C, as shown in Fig. 4. Let the absolute refractive index of the lens materials be μ_2; and the lens be surrounded on all sides by air or any other rarer medium of refractive index μ_1. The centers of curvatures of the two refracting surfaces of the lens are C_1 and C_2. Let the radius of the curvature of the surfaces be R_1 and R_2 respectively. Consider a point object 'O' situated on the principal axis of the lens.

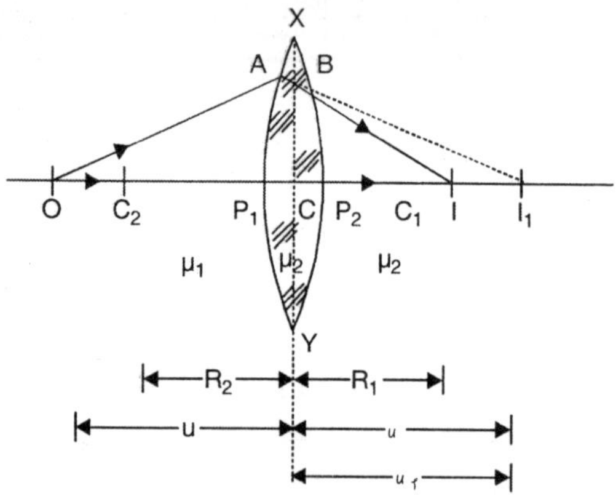

Fig. 4

A ray of light leaving the axial point-object O and travelling along the principal axis passes without deviation. Another ray OA strikes the first surface at A and is refracted in a direction BI_1.

The ray is further refracted by the second surface in a direction BI and meets the ray along the principal axis at I. Therefore, I is the final point-image of the object O formed after refraction by the two surfaces of the lens.

By considering refraction at spherical surfaces we can derive an equation that describes image formation by a lens. The basic idea is that the image formed by the first refracting surface acts as a *virtual object* for the second refracting surface. We can apply *Gauss formula*

$$\frac{\mu_1}{u} + \frac{\mu_2}{\upsilon} = \frac{\mu_2 - \mu_1}{R} \qquad ...(1)$$

for refraction at the first surface.

The first surface, XAP_1Y, forms a *real* point-image, I_1. We replace υ by υ_1 and R by R_1 in eqn. (1) and apply the sign convention. We see that u is –ve. Therefore, the equation (1) is written for the first surface as

$$\frac{\mu_2}{\upsilon_1} - \frac{\mu_1}{u} = \frac{\mu_2 - \mu_1}{R_1} \qquad ...(2)$$

In the Fig. 4, the light leaving the first surface would reach I_1 if the second surface XBP_2Y did not intervene. When the second surface is present, the light does not reach I_1. I_1 becomes the virtual object for the second surface and the second surface forms the image at I, which is the final image. The position of the final image is again found using Gauss formula using $u = \upsilon$ and $R = R_2$. Thus, we get

$$\frac{\mu_2}{\upsilon_1} + \frac{\mu_1}{\upsilon} = \frac{\mu_2 - \mu_1}{R_2}$$

Applying the sign convention, we get

$$\frac{\mu_2}{\upsilon_1} + \frac{\mu_1}{\upsilon} = \frac{\mu_2 - \mu_1}{R_2} \qquad ...(3)$$

Adding equations (2) and (3), we obtain

$$\mu_1\left(\frac{1}{\upsilon} - \frac{1}{u}\right) = (\mu_2 - \mu_1)\left[\frac{1}{R_1} - \frac{1}{R_2}\right]$$

$$\left(\frac{1}{\upsilon} - \frac{1}{u}\right) = \left(\frac{\mu_2 - \mu_1}{\mu_1}\right)\left[\frac{1}{R_1} - \frac{1}{R_2}\right]$$

$$\frac{1}{\upsilon} - \frac{1}{u} = \left(\frac{\mu_2}{\mu_1} - 1\right)\left[\frac{1}{R_1} - \frac{1}{R_2}\right] \qquad ....(4)$$

For air, $\mu_1 = 1$. Designating μ_2 as μ, we get

$$\frac{1}{\upsilon} - \frac{1}{u} = (\mu - 1)\left[\frac{1}{R_1} - \frac{1}{R_2}\right] \qquad ...(5)$$

Equation (5) relates the image distance l) of a thin lens to the object distance u and to the thin lens properties namely, refractive index and the radii of curvature. It is to be noted that a lens will not only focus from O to I but between any other pairs of points, as long as the points satisfy the relation (5).

Notes :

1. The above formula is valid *only* for nearly paraxial rays and strictly a *thin* lens.

2. The initial equation (2) is a *numerical relationship* between the various distances for a particular type of lens surface. We then adopted a sign convention for the distances, which enabled us derive an *algebraic relationship.* Such an algebraic relation (5) obtained with the sign convention will serve for all types of lenses used in all possible ways. Any sign convention that is different than that adopted by us is likely to give a different algebraic relationship.
3. u, υ, f and R are algebraic quantities each having a sign, numerical value and a unit. While solving the numerical problems, we have to assign the proper signs and values for these three parameters.

LENS MAKER'S EQUATION

If the object is at infinity, the image is formed at the *principal focus* of the lens. When $u = \infty$ $\frac{1}{u} = 0$ and $\upsilon = f$. Eqn. (5) becomes

$$\frac{1}{f} - \frac{1}{\infty} = (\mu - 1)\left[\frac{1}{R_1} - \frac{1}{R_2}\right]$$

$$\therefore \qquad \frac{1}{f} = (\mu - 1)\left[\frac{1}{R_1} - \frac{1}{R_2}\right] \qquad ...(6)$$

Eqn. (6) is known as the *lens makers' formula*, since it enables one to calculate/from the known properties of the lens. It can also be used to determine the values of R_1 and R_2 needed for a desired focal length of a lens of a given index of refraction.

If the lens is turned around, R_1 and R_2 are interchanged and the sign of each is reversed. Consequently, there is no change in f. Therefore, *for a thin lens, the focal length is independent of* the order of the surfaces.

It is clear from the lens makers' formula that to get a short focal length, the lens has to have a surface of smaller radius of curvature R and be made of a material with a high refractive index.

Comparing equations (5) and (6), we see that

$$\frac{1}{\upsilon} - \frac{1}{u} = \frac{1}{f} \qquad ...(7)$$

The above equation is known as the *Gauss' formula for a lens.* This is a more useful formula than eqn. (5), because if we know the focal length of the lens, we can relate the position of the object and its image without necessarily knowing the index of refraction or the radii of the lens.

Notes :

1. The μ in eqn. (6) stands for the refractive index of the lens material with respect to the surrounding medium. If the lens is placed in a medium such as water, then the value of μ should be the refractive index of glass with respect to water. That is,

$$\mu = \mu_{wg}$$

or $$\mu = \mu_g/\mu_w.$$

2. Another sign convention is there in which the distances of the *real* objects or images from the lens would be considered as *positive* and those of *virtual* images would be regarded as *negative*. Though, that sign convention is simple, it is found inadequate in many respects. When that sign convention is used, we get the following formula for the focal length of a lens. $\frac{1}{u} + \frac{1}{\upsilon} = \frac{1}{f}$

CONSTRUCTION OF THE IMAGE USING CARDINAL POINTS

From the knowledge of the cardinal points of an optical system, the image corresponding to any object placed on the principal axis of the system can be constructed. It is not necessary to know the position and curvatures of the refracting surfaces or the nature of the intermediate media. Only knowledge of cardinal points and cardinal planes is sufficient.

Let F_1, F_2 be the principal foci, P_1, P_2 the principal points and N_1, N_2 the nodal points of the optical system. AB is a linear object on the axis. In order to find the image of the point A we make the following construction.

(1) A ray AH_1 is drawn parallel to the axis touching the first principal plane at H_1. The conjugate ray will proceed from H_2 a point on the second principal plane such that $H_2P_2 = H_1P_1$ and will pass through the second principal focus F_2.

(2) A second ray AF_1 is drawn passing through the first principal focus F_1 and touching the first principal plane at K_1. Its conjugate ray will proceed from K_2 such that $K_2P_2 = K_1P_1$ and it will be parallel to the axis.

(3) A third ray $A_1T_1N_1$ is drawn which is directed toward the first nodal point N_1. This ray passes after refraction through N_2 in a direction parallel to AN_1.

The point of intersection of any of the above two refracted rays will give the image of A.

Let it be A_1. From A_1, if a perpendicular is drawn on to the axis, it gives the image A_1B_1 of the object AB.

NEWTON'S FORMULA

We have Δ^{les} ABF_1 and $F_1K_1P_1$ are similar.

$$\therefore \qquad \frac{K_1P_1}{AB} = \frac{P_1F_1}{BF_1}$$

But $K_1P_1 = A_1B_1$

$$\therefore \qquad \frac{A_1B_1}{AB} = \frac{f_1}{x_1} \qquad ...(1)$$

Further, $\Delta^{les}A_1B_1F_2$ and $H_2P_2F_2$ are similar.

$$\frac{A_1B_1}{H_2P_2} = \frac{B_1F_2}{P_2F_2}$$

But, $H_2P_2 = AB$

$$\therefore \qquad \frac{A_1B_1}{AB} = \frac{x_2}{f_2} \qquad ...(2)$$

From equations (2) and (3), we get

$$\frac{h_2}{h_1} = \frac{f_1}{x_1} = \frac{x_2}{f_2} \qquad ...(3)$$

$$\text{or} \qquad x_1x_2 = f_1f_2 \qquad ...(4)$$

This is the *Newton's formula*. In the foregoing discussion, the distances of the image and the object have been measured from their respective foci. But it is sometimes convenient to measure the conjugate distances from the principal points.

POSITIONS OF THE PRINCIPAL FOCI

Each individual surface of the lens has its own focal points and planes and the lens as a whole has its own *pair* of focal points and focal planes. The focal points and focal planes of the lens are known as principal focal points and principal focal planes. We are interested in knowing the locations of these principal focal points and principal focal planes. They are obtained as follows.

(i) If a point object is placed on the principal axis such that the rays refracted by the lens are parallel to the axis, then the position of the point object is called the *first principal focus* F_1 of the lens (Fig. 5a). The distance at the first principal focus from the optical center C of the lens is called the *first principal focal length*, f_r We can find f_1 as follows.

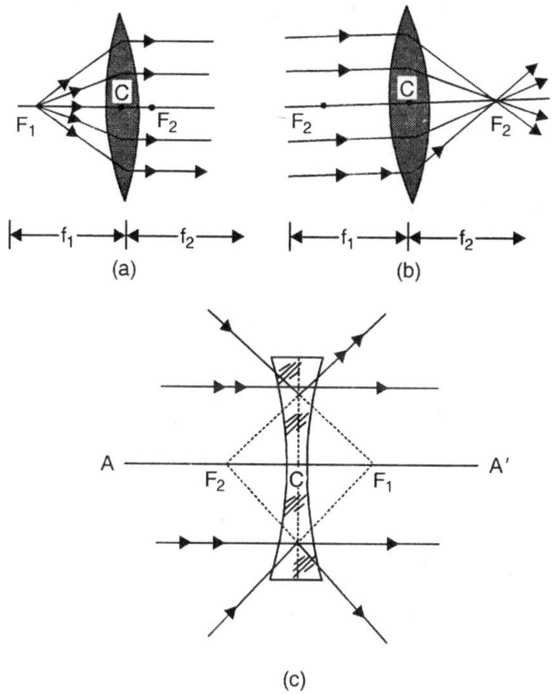

Fig. 5

Using $u = f_1$, and $\upsilon = \infty$ into eqn. (4.6), we get

$$\frac{1}{\infty} - \left(-\frac{1}{f_1}\right) = (\mu - 1)\left[\frac{1}{R_1} - \frac{1}{R_2}\right]$$

or

$$\frac{1}{f_1} = (\mu - 1)\left[\frac{1}{R_1} - \frac{1}{R_2}\right] \qquad ...(8)$$

The plane perpendicular to the axis and passing through the first focal point is known as the *first principal focal plane.*

(ii) If the object is situated at infinity, the position of the image on the axis is known as the second principal focus F_2 (Fig. 5b). The distance of the second principal focus from the optical center C is called the *second principal focal length,* f_2.

Using $u = \infty$ and $\upsilon = f_2$ into eqn. (5), we get

$$\frac{1}{f_1} - \frac{1}{\infty} = (\mu - 1)\left[\frac{1}{R_1} - \frac{1}{R_2}\right]$$

$$\frac{1}{f_2} = (\mu - 1)\left[\frac{1}{R_1} - \frac{1}{R_2}\right] \quad ...(9)$$

The plane perpendicular to the axis and passing through the second focal point is known as the *second principal focal plane.*

It follows from eqn. (8) and (9) that

$$f_1 = f_2 \quad ...(10)$$

Thus, *every thin lens in air has two focal points (F_1 and F_2), one on each side of the lens and equidistant from the centre.*

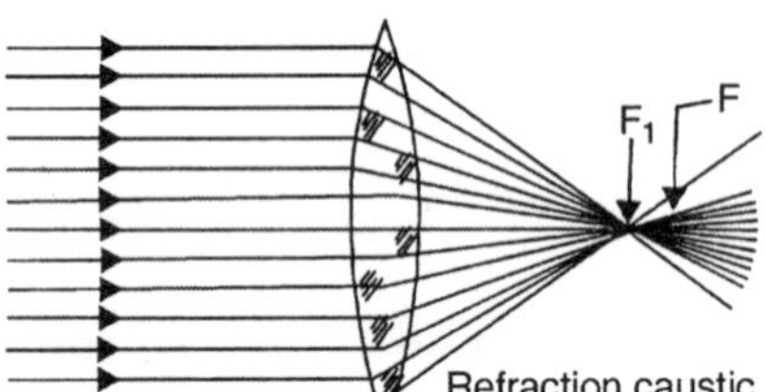

Fig. 6.

It will be seen that the second focal length (f_2) of a converging lens is positive and the first (f_1 negative, while for a diverging lens the reverse is true (Fig. 5c).

The two focal lengths of thin lens in air are numerically equal. Frequently, one refers simply to "focal length" of a lens; it will be assumed that this always refers to the *second focal length* so that the focal length of a converging lens is positive, while for a diverging lens it is negative.

Note : For lenses having a large diameter or aperture, the image of a point is not a point. Incident rays that are parallel to the principal axis intersect at different points, depending on their distance from that axis. The refracted rays intersect over a conical surface called the *refraction caustic.*

LENS EQUATION OF THE NEWTON'S

The lens-maker's formula (6) tells the lens-maker what is needed to know to grind a lens of a given power: the refractive index of the glass and the two radii of curvature.

Now we obtain a lens equation for the practical applications. It relates the focal length to the object and image distances. This is also known as *lens users' formula or Newton's lens equation.*

Let h_1 be the height of the object and h_2 be the image height. From the similar triangles to the left of the lens we find that

To the right of the lens we have

$$\frac{h_1}{f_2} = \frac{h_2}{x_2}$$

Rearranging both equations and combining them by eliminating $\frac{h_1}{h_2}$ gives

$$\frac{x_1}{f_1} = \frac{f_2}{x_2}$$

or $$x_1 x_2 = f_1 f_2. \qquad \text{...(11)}$$

This is known as *Newton's lens equation.* The terms x, and x, are extra focal object and image distances respectively. They are *measured from the foci rather than from the lens.* Consequently, Newton's equation can be used with both "thin" and "thick" lenses.

When the medium is the same on both sides of the lens, the eqn. (11) reduces to

$$x_1 x_2 = f^2 \qquad \text{...(12)}$$

Note: In the Gaussian formula the object and image distances are measured from the lens centre, while in the Newton's formula they are measured from focal points.

LONGITUDINAL AND LATERAL MAGNIFICATION

The *longitudinal magnification* is defined as the ratio of an infinitesimal axial length in the region of the image to the corresponding length in the region of the object. Thus, if dx_i, and dx_o are the extensions of the object and the image along the principal axis respectively, then the longitudinal magnification of the lens is given by

$$m_L = \frac{dx_i}{dx_o} \qquad ...(13)$$

Differentiating eqn. (12), we get

$$m_L = -\frac{f^2}{x_o^2} = -m^2 \qquad ...(14)$$

Lenses are used to form a magnified image of an object. Magnification is defined as

$$\text{Magnification, } m = \frac{\text{size of the image}}{\text{size of the object}}$$

The image of a three-dimensional object will occupy a three- dimensional region of space. The lens can affect the transverse, longitudinal and angular dimensions of the image. Since the image and object relationship is expressible in terms of axial, lateral and angular distances, we distinguish three types of magnification, namely *lateral magnification, longitudinal magnification and angular magnification.*

Lateral or *transverse magnification* of a lens is defined as the ratio of the length of the image to the length of the object size, both the lengths being measured perpendicular to the principal axis. Thus,

$$m = \frac{h_2}{h_1} \qquad ...(15)$$

According to sign convention, the distances above the principal axis of the lens are taken positive and those below the axis are negative. Hence, the lateral magnification is positive for an erect image and negative for an inverted image.

If the refractive indices on either side of the lens are the same, then

$$\frac{h_1}{u} = \frac{h_2}{\upsilon}$$

$$\therefore \qquad m \frac{\upsilon}{u} \qquad ...(16)$$

The lateral magnification corresponding to Newton's formula may be written as

$$m = \frac{h_2}{h_1} = \frac{f_1}{x_1} = \frac{x_2}{f_2} \qquad ...(17)$$

ANGULAR MAGNIFICATION

Angular magnification is defined as the ratio of slopes of emergent ray and conjugate incident ray with the principal axis. If the incident ray and the conjugate emergent ray make angles θ_1 and θ_2 with the principal axis respectively, then the angular magnification of the lens is given by

$$\gamma = \frac{\tan \theta_2}{\tan \theta_1} \qquad ...(18)$$

DISPLACEMENT OF LENS WHEN OBJECT AND SCREEN ARE FIXED

Let us consider a linear object, O, placed in front of a convex lens. A, which gives rise to an image on the screen at I. Since the image distance AI (= υ) is greater than the object distance AO (= u), the image is larger than the object. If the object and the screen are kept fixed at O and I respectively, and the lens is moved from A to position B, then another clear image is obtained. This time the image is smaller than the object. Since O and I are conjugate, OB = IA and IB = OA. If we designate the displacement AB of the lens by d and the constant distance O*l* by *l*, then OA + BI = l – d.

$$\therefore \qquad OA = (l - d)/2.$$

And AI = AB + BI = OA + AB = $(l - d)/2 + d = (l + d)/2$.

But OA = u and AI = υ for the lens position A. Substituting for OA and AI in the lens formula, we get

$$\frac{2}{l + d} - \frac{2}{l - d} = \frac{1}{f}$$

$$\therefore \qquad \frac{4l}{l^2 - d^2} = \frac{1}{f}$$

$$\therefore \qquad f = \frac{l^2 - d^2}{4l} \qquad ...(19)$$

Thus, if the displacement d of the lens and the distance *l* between the object and the screen are measured, the focal length f of the lens can be found from the above equation (19). This gives us a very useful method of measuring the focal length of a lens.

(ii) *Magnification* : When the lens in position A, the lateral magnification, m_1 of the object is

$$m_1 \frac{\upsilon}{u} = \frac{AI}{OA}$$

As AI > OA, the image, h_1, is bigger than the object, h.

$$\frac{h_1}{h} = \frac{AI}{OA} \qquad ...(20)$$

When the lens is in position B, the image is smaller than the object. The lateral magnification is given by

$$m_2 = \frac{BI}{OB} \therefore \frac{h_1}{h} = \frac{BI}{OB} \qquad ...(21)$$

As AI = OB and OA = BI. we get from eqn. (20) and (21)

$$\frac{h}{h_1} = \frac{h_2}{h}$$

$$\therefore \qquad h_2 = h_1h_2 \qquad ...(22)$$

$$h = \sqrt{h_1h_2}$$

It means that the size of the object is equal to the geometric mean of the two images. This method of measuring the size of the object is useful when the object is inaccessible.

OBJECT AND REAL IMAGE

We find that it is not always possible for a convex lens to form a real image of the object on a screen, although the object and the screen may both may be at a greater distance from the lens than its focal length. A convex lens forms a real or a virtual image depending on the distance of the object from the lens.

We now show that the distance between an object and a screen must be equal to or greater than '4f' if a real image is to be formed with a convex lens.

Suppose I is the real image of a point object O in a converging lens. If the image distance is x, and the distance OI = d, the object distance = [d – x] (Fig. 7). Thus, $\upsilon = x$, and $u = (d - x)$. Substituting these in the lens equation, we get

$$\frac{1}{x} + \frac{1}{d-x} = \frac{1}{f}$$

$$\therefore \quad \frac{d}{x(d-x)} = \frac{1}{x}$$

$$\therefore \quad x^2 - \; xd = fd = 0.$$

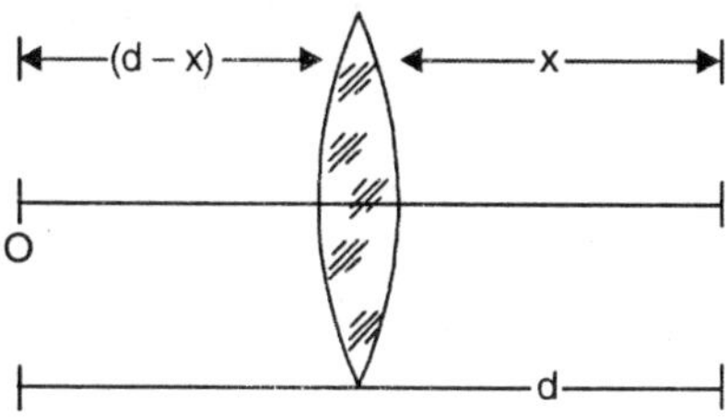

Fig. 7.

If the image is to be real, the roots of this quadratic equation for x must be real.

$$\therefore \quad d^2 - 4fd > 0 \qquad \text{or} \quad d > 4f. \qquad ...(23)$$

Hence, the *minimum* distance between an object and its real image for a convex lens must be 4f where f is the focal length of the lens.

EQUIVALENT FOCAL LENGTH OF TWO THIN LENSES

When two thin lenses are arranged coaxially, the image formed by the first lens system becomes the object for a second lens system and the two systems act as a single optical system forming the final image from the original object.

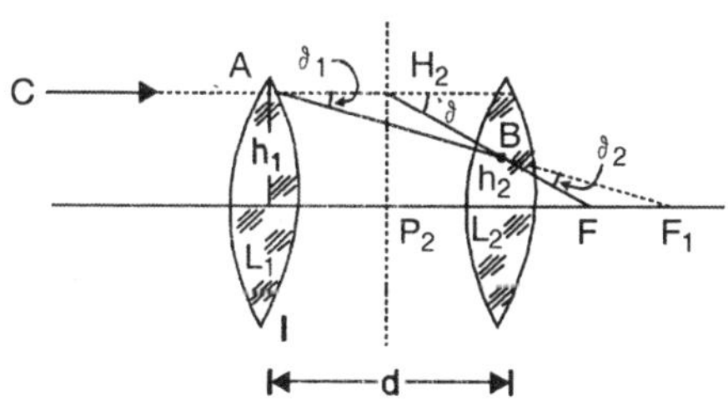

Fig. 8.

Let us consider a simple optical system that consists of two thin lenses L_1 and L_2 placed on a common axis and separated by a distance d, as shown in Fig. 8. The lenses are separated by a distance and have focal lengths f_1 and f_2. We are interested to know how the combination works. We find that two lenses, separated by a finite distance, can be replaced by a single thin lens called an *equivalent lens*.

The equivalent lens, when placed at a suitable fixed point, will produce an image of the same size as that produced by the combination of the two lenses. The focal length of equivalent lens is called equivalent focal length. We now derive an expression for the equivalent focal length, f, of the combination of two lenses. Let a ray CA of monochromatic light parallel to the principal axis be incident on the first lens L_1 at a height h_1 above the axis. The ray CA is deviated through an angle δ_1 by the lens L_1. The incident ray CA, after refraction, is directed toward F_1, which is the *second principal focus of lens* L_1. Then the deviation produced by the first lens is given by

DEVIATION BY A THIN LENS

A lens may be considered to be made up of a large number of prisms placed one above the other. As the action of the lens is to-deviate the incident rays of light, it is necessary to find the deviation produced by a particular section of the lens. Let a ray of monochromatic light parallel to the principal axis be incident on a thin lens at a height h above the axis and let/be the focal length of the lens. As the ray is parallel to the principal axis, after refraction it will pass through the second focus, F_2 (Fig. 9a).

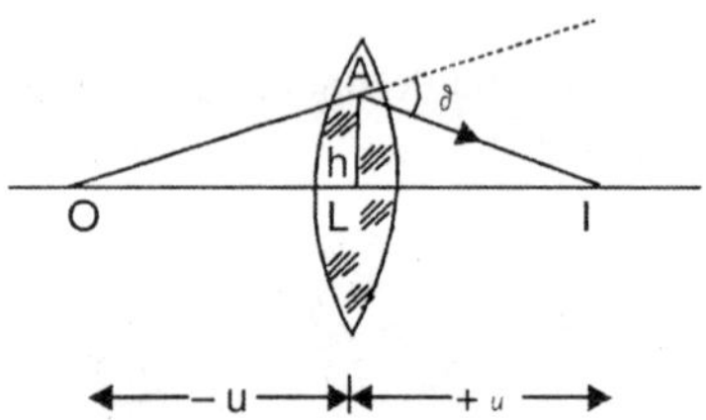

Fig. 9

The deviation suffered by the ray is given by

$$\tan \delta = \frac{h}{f}$$

In the paraxial region 5 being small, $\tan \delta = \delta$.

$$\therefore \qquad \delta = \frac{h}{f}$$

Next, consider a luminous point object O and its corresponding image I (Fig. 4.13b). The deviation suffered by the ray OA incident at A is given by

$$\delta = \angle AOL + \angle AIL$$

$$\delta = \frac{h}{-u} + \frac{h}{+\upsilon} = h\left[\frac{1}{\upsilon} - \frac{1}{u}\right] = +h\left[\frac{1}{f}\right]$$

$$\therefore \qquad \delta = \frac{h}{f} \qquad ...(24)$$

This shows that the deviation produced by a lens is *independent* of the position of the object.

Eqn. (24) is valid for all rays incident at the same point A of a lens, whatever may be the direction of incidence. All the rays suffer the same deviation in refraction through the lens.

Power

The *power* of a lens is the measure of its ability to produce convergence of a parallel beam of light. A convex lens of large focal length produces a small converging effect and a convex lens of small focal length produces a large converging effect. Due to this reason, the power of a *convex lens* is taken as *positive* and a convex lens of small focal length has high power. On the other hand, a *concave lens* produces divergence. Therefore, its power is taken as *negative*.

The unit in which the power of a lens is measured is called a *diopter* (D). A convex lens of focal length one metre has a power = +1 diopter and a convex lens of focal length 2 m has a power = + 0.5 diopter.

$$\text{Power} = \frac{1}{\text{Focal length in metres}} \qquad ...(25)$$

The concept power is useful because it allows us to work out the effective focal length of a combination of lenses very easily. The power of a pair of lenses, of focal lengths f_1 and f_2, placed in contact is simply the sum of their individual powers.

$$\frac{1}{f} = \frac{1}{f_1} + \frac{1}{f_2}$$

$$\therefore \qquad P = P_1 + P_2 \qquad ...(26)$$

where P_1 and P_2 are the powers of the two lenses and P is the equivalent power.

GAUSS'S GENERAL THEORY OF IMAGE FORMATION

Gauss's theory deals with ideal image formation, by which we mean cases where there is a point-for-point, line-for-line and plane-for-plane relationship between an object and its image.

With the various refracting or reflecting surfaces coaxial, the common axis is taken as x-axis. Due to symmetry about this axis, it is enough to deal with any one plane through the x-axis, and the transverse distances give y.

With arbitrary choice of origins for the object and image space, let (x, y) be the co-ordinates of an object point and (x', y') those for its image. Then the most general linear relationships of x', y' with x, y would be:

$$x' = \frac{a_1x + b_1y + d_1}{ax + by + d}$$

$$y' = \frac{a_2x + b_2y + d_2}{ax + by + d} \qquad ...(1)$$

We state this without a rigorous mathematical proof. We can see that out of the nine constants involved, five are redundant. Firstly, if y changes sign, x' should remain unaffected. This condition makes

$$b = b_1 = 0.$$

Again, if y changes sign, y' should only change in sign, not in magnitude. This makes $a_2 = d_2 = 0$. Finally dividing all coefficients by a, and expressing the new values by the corresponding capital letters, we get

$$x' = \frac{A_1x + D_1}{x+D}, \quad y' = \frac{B_2y}{x+D} \qquad ...(2)$$

Thus, we reach the very important conclusion that in *any co-axial symmetrical system forming ideal images the co-ordinates x', y' of an image point corresponding to the object point x, y are uniquely determined by four and only four independent constants.*

In actual practice, two of these constants are used to Specify the origins for object space and image space on the x-axis for measurement of distances.

Then only two other constants specify the object to image conjugate relations uniquely. There are several alternative ways of choosing the origins, and we shall discuss two of them in the following sections.

HELMHOLTZ'S LAWS

These laws relate to slope angles of conjugate rays from axial source and image points and the transverse magnification when point-to-point and plane-to-plane, correspondence between source and its image exists. The involved surfaces need not be spherical; they could be corrected to ensure the one-to-one correspondence described above.

The P and Q are conjugate axial points. Rays PA and PO are refracted as AB and OC, intersecting at Q to give the image point. The slope angles are related through

$$\frac{\tan \theta_1}{\tan(-\theta_2)} = \frac{AO'/PO'}{AO'/QO'} = \frac{\upsilon}{-u} \qquad ...(9)$$

The only assumption involved is that O' is close enough to O to permit taking PO' = PO and QO' = QO. This equation says that if all the rays at different θ_1 after reduction, meet at a common point Q (fixed υ), then the ratio of tangents of the slopes remains constant. If P' is a transverse point from P distant y_1, and a ray P'O refracts as OD, its intersection with a line transverse at Q would give the image Q' since one-to-one point and plane correspondence is assumed to hold. The transverse distances y_1 are much smaller than u or υ, and hence the approximation $\sin \theta_i = \theta_i$ is valid. If n_1 and n_2 are the refractive indices of media on the left and right, we get

$$\frac{n_1}{n_2} = \frac{\sin \theta_2}{\sin \theta_1} = \frac{\theta_2}{\theta_1} = \frac{-y_2/\upsilon}{y_1/(-u)} = \frac{y_2}{y_1}.\frac{u}{\upsilon} \qquad ...(10)$$

Eliminating $\frac{\upsilon}{u}$ between Eqs. (9) and (10) leads to

$$n_1 y_1 \tan \theta_1 = n_2 y_2 \tan \theta_2 \qquad ...(11)$$

This is known as *Lagrange's equation for magnification.* It is of especial significance in microscopic images where y_1 is always small but θ_1 values are quite large. We repeat that the deduction does not relate to spherical or any special surfaces, but to conditions under which one-to-one conjugate relation holds over the entire range of θ_1 values.

Eqn. (11) is of very general application. Since longitudinal distances are not involved and only the slopes come in, it can be applied to a succession of coaxial surfaces irrespective of their separations. Let n_1, n_2 ... n_p be refractive indices of successive media, $\theta_1, \theta_2 ... \theta_p$, the slopes of a given ray in these media and $y_1, y_2 ... y_p$ heights of the object and Successive images Then we may apply Eqn. (11) in succession to get

$$n_1 y_1 \tan \theta_1 = n_p y_p \tan \theta_p.$$

This is the *Lagrange—Helmholtz Law of Magnification.*

CARDINAL POINTS

In the case of refraction through a thin lens, the thickness of the lens has been neglected in calculating the various formulae. It is then immaterial from which point of the lens the distances are measured. But we cannot apply the above approximation for an optical system consisting of a combination of lenses (or for a *thick lens*). One way of calculating the position and size of the image formed by an optical system is to consider refraction at each surface of a lens successively, but the method proves to be more tedious. In 1841, Gauss showed that any number of coaxial lenses could be treated as a single unit, without the necessity of treating the single surfaces of lenses separately. The lens makers' formula can be applied to the system provided the distances are measured from two *hypothetical parallel planes*, fixed with reference to the refracting system. The points of intersection of these planes with the axis are called the *principal points* or *Gauss points*. In fact there are six points in all, which characterize an optical system. They are:

(i) two focal points, (ii) two principal points, and

(iii) two nodal points.

These six points are known as *cardinal points* of an optical system. The planes passing through these points and which are perpendicular to the principal axis are known as *cardinal planes*. The cardinal points and cardinal planes are intrinsic properties of a particular optical system and determine the image forming properties of the system. If these are known, one can find the image of any object without making a detailed study of the passage of the rays through the system. It is not necessary to consider the refraction of the rays at the various surfaces.

We describe here how to locate the, cardinal points and planes for a coaxial optical system.

CARDINAL POINTS OF AN OPTICAL SYSTEM

In any system of lenses or mirrors or both, which forms ideal images together are called the *cardinal points* of the system. If these are given then the image point f corresponding to any object point P can be determined either from eqn. (3 and 4) geometrical drawing.

Now we note that the conjugate of point C must lie in the second principal plane such that H'C' = HC. This gives us the point C. We also know that since ray PFC passes through the first focus F, its conjugate must be parallel to the principal axis. So from C we draw a line CM parallel to the principal axis.

Next, from P we draw PD parallel to the principal axis, meeting the first principal plane at D. The conjugate of D is D', obtained by the rule that it must be in the second principal plane, with H'D' = HD. Further, since ray PD is parallel to the principal axis, its conjugate must pass through f'. So from D' we draw a ray D'N passing through F'. Its intersection with CD gives us F.

Corresponding to the two incident rays PC and PD, the emergent rays meet at P'. Since the system is assumed to give ideal images, all rays starting from P must on emergence pass through P'. Thus, P' is the conjugate of object point P. It may be noted that this tracing required all the *four* points F,F', H,H'; less will not do and more information is not needed.

ACTUAL SYSTEMS

The general treatment given so far does not tell us where F, F' lie in any actual system, and whether focal lengths f and f' bear any relation to one another. To determine these inter-relationships we shall require the application of the *actual* laws of refraction (or reflection) in some generalized form.

However, before we do that it will be useful to examine some familiar simple cases like those of single refraction and a thin lens to discover if they indicate some simple relationships.

(a) *Refraction at a Spherical Surface :* OA represents the spherical surface of separation of two media of refractive indices n_1 and n_2. C is the centre of the surface, P is an object point and PC is the principal axis. It meets the spherical surface at O, which gives the pole. For a ray PA, the normal is CAN, and the refracted ray is AM, meeting the principal axis at Q, which is the image of P.

Using Snell's law $n_1 \sin i = n_2 \sin r$ (with first order approximation $i = 2_n r$) and the geometry of one deduces

$$n_1 (\theta_0 + \theta_1) = n_2 (\theta_o - \theta_2)$$

$$\Rightarrow \qquad n_1\,\theta_1 + n_2\,\theta_2 = (n_2 - \theta_1)\,\theta_o$$

or
$$\frac{n_2}{\upsilon} - \frac{n_1}{u} = \frac{n_2 - n_1}{R} \qquad ...(1)$$

Now, if we put $\upsilon = \infty$, the corresponding u is, by definition, the *focal length of object space*, f. If we put $u = \infty$, the corresponding υ is, by definition, the *focal length of image space f'*. We thus get

$$f = -\frac{n_1 R}{n_2 - n_1}; \qquad f' = +\frac{n_2 R}{n_2 - n_1} \qquad ...(2)$$

We note that f and f' are related by

$$\frac{f'}{f} = -\frac{n_2}{n_1} \qquad ...(3)$$

Relations (4) give f and f' for the particular system in terms of n values and R. But Eq. (3) may have a more *general validity*.

(b) *Thin Lens in Any Medium* : Consider a thin lens of radii of curvature R_1, R_2 and refractive index n. For generality, let the initial and final media have refractive indices n_1 and n_2 respectively. Then applying Eqn. (1) in two steps, one gets

$$\frac{n_2}{\upsilon} - \frac{n_1}{u} = n\left(\frac{1}{R_1} - \frac{1}{R_2}\right) + \left(\frac{n_2}{R_2} - \frac{n_1}{R_1}\right) = \frac{1}{G}\text{(say)}, \qquad ...(4)$$

where G is a constant of the system having the dimensions of length. On applying the definition $u \equiv f$ when $\upsilon = \infty$ and $\upsilon = f'$ when $u = \infty$, we get

$$f = n_1 G \quad \text{and} \quad f' = +\,n_2 G \qquad ...(5)$$

so that again

$$\frac{f'}{f} = -\frac{n_2}{n_1}$$

SOLVED EXAMPLES

Example 1:

A convex lens of thickness 4 cm has radii of curvature 6 cm and 10 cm. Find the focal length and the positions of the focal points and the principal points. The refractive index of lens material is $\mu = 1.5$.

Solution:

Here $\qquad R_1 = 6$ cm,

$$R_2 = -\ 10 \text{ cm}$$

and $t = 4$ cm.

$$f = \frac{-\mu R_1 R_2}{\mu(\mu-1)(R_1 - R_2) - (\mu-1)^2 t}$$

$$= \frac{-1.5 \times 6\text{cm} \times (-10\text{cm})}{1.5(1.5-1)(6\text{cm}+10\text{cm}) - (1.5-1)^2 \times 4\text{cm}}$$

$$f = 8.18 \text{ cm.}$$

$$\alpha = \frac{R_1 t}{\mu(R_1 - R_2) - t(\mu - t)}$$

$$= \frac{6\text{cm} \times 4\text{cm}}{1.5[6\text{cm} - (-10\text{cm})] - 4\text{cm}(1.5-1)}$$

$$= 1.09 \text{ cm.}$$

$$\beta = \frac{R_2 t}{\mu(R_1 - R_2) - t(\mu - 1)}$$

$$= \frac{4\text{cm}(-10\text{cm})}{1.5[6\text{cm} - (-10\text{cm})] - 4\text{cm}(1.5-1)}$$

$$= -\ 1.82 \text{ cm.}$$

P_1 and P_2 show the positions of the principal points and F_1 and F_2 show the positions of the focal points. As the medium on the two sides of the lens is the same, the nodal points coincide with the principal points.

Example 2:

Deduce the positions of cardinal points for a plano convex lens of refractive index 1.6, the radius of curvature of the curved surface being 6.0 cm and thickness 2.0 cm.

Solution:

Let light enter from the curved surface. We then have $R_1 = +\ 6$ cm, $t = 2$ cm, $R_2 = \infty$. Whence,

$$\frac{1}{f'} = (1.6 - 1)\ \frac{1}{f'};$$

$$\Rightarrow \qquad f' = \frac{6}{0.6} = +\ 10.0 \text{ cm}$$

$$\alpha = -f'\frac{(n-1)t}{nR_1}$$

$$= -10x\frac{0.6\times2}{1.6\times6} = -1.25 \text{ cm.}$$

$$\alpha = -f'\frac{(n-1)t}{nR_2} = 0 \text{ cm} \qquad (\because R_2 = \infty)$$

Tthe results. First, a' = –1.25 cm gives *H' to the left of O'*. Next use f' = + 10.0 cm, which gives F' at 10.0 cm, *to the right of H'*. Since $\alpha = 0$, the point H falls on the curved surface.

Finally, we have f = – f' = – 10.0 cm.

So F lies 10.0 cm *to the left of H.*

The same results come if light enters from the right so far as the positions are concerned. Only Symbols F, H and F', H' get interchanged.

Example 3:

The radii of curvature of a convex meniscus lens are 15 cm and 10cm and its thickness is 2 cm. Calculate its focal length and the positions of cardinal points. If an object is placed at a distance of 100cm from this lens, calculate the position and magnification of the image.

Solution:

A meniscus lens has both the surfaces curved in the same direction. In a convex meniscus lens $R_1 > R_2$.

Given $R_1 = 15$,

$R_2 = 10$ cm,

t = 2 cm and $\mu = 1.5$.

Example 4:

A sunshine record globe of 30 cm diameter is made of glass ($\mu = 1.5$). A ray enters the globe parallel to the axis. Find the position where the ray crosses the axis.

Solution:

A sunshine recorder globe is a sphere of glass. The radius of curvature of the first surface is + 15 cm and u = ∞.

$$\frac{\mu}{v_1} - \frac{1}{u} = \frac{(\mu - 1)}{R}$$

$$\frac{1.5}{v_1} - \frac{1}{\infty} = \frac{(1.56 - 1)}{15\text{cm}}$$

$\Rightarrow$ $v_1 = 45$ cm

$\therefore$ $P_1C = 45$ cm.

For the second surface, point C is the virtual object.

$\therefore$ $CP_2 = 45$ cm $-$ 30 cm = 15 cm.

At the second surface, the ray asses from a denser to a rarer medium and R = $-$ 15 cm.

$$\frac{1/\mu}{v} - \frac{1}{u} = \frac{(1/\mu - 1)}{R}$$

$$\frac{1/1.5}{v} - \frac{1}{15\text{cm}} = \frac{(1/1.5 - 1)}{-15\text{cm}}$$

$\therefore$ v = 7.5 cm

Therefore, the ray crosses the axis at a point D which is 5 cm from the point P_2 of the globe.

Example 5:

A convex meniscus lens has radii of curvature 3.0 and 2 cm, and thickness 1.0 cm. If n = 1.50, deduce the positions of cardinal points.

Solution:

A 'meniscus' lens is one which has both surfaces curved in the same direction. If the face on the concave side is less curved (larger R) then the lens is convex.

With light travelling left to right, $R_1 = -3$, $R_2 = -2$, t = 1 (all in cm). We have, with n = 1.5.

$$\frac{1}{f'} = (0.5)\left(\frac{1}{-3} - \frac{1}{-2}\right) + \frac{(0.5)^2 \times 1}{1.5(-3)(-2)}$$

$\therefore$ f' = + 9.0 cm (check), f = $-$ 9.0 cm.

$$\alpha' = -9.0 \times \frac{(1.5 - 1) \times 1}{1.5 \times (-3)} = +1.0 \text{ cm}$$

$$\alpha = -9.0 \times \frac{(1.5-1)\times 1}{1.5\times(-2)} = +1.5 \text{ cm}$$

We plot H' at 1.0 cm *to the right of O'* and H at 1.5 cm to the right of O. Now F' lies at the right of H', and F at 9.0 cm to the left of H. Note that both the principal planes fall outside the lens and on the same side.

Example 6:

Two converging lenses of focal lengths 20 and 10 cm are placed 25 cm apart in air. Find and plot the positions of H,H', F and F'. Also find the position and magnification of the image of an object placed 10 cm in front of the first lens.

Solution:

$$\nabla = f_1' - f_2' - t = 20 + 10 - 25 = 5 \text{ cm}$$

$$\therefore \quad f' = \frac{f_1' f_2'}{\nabla} = \frac{20\times 10}{5} = 40 \text{ cm};$$

$$f = -f' = -40 \text{ cm}$$

$$a' = -f' \frac{t}{f_1'} = -50 \text{ cm}$$

$$a = f' \frac{t}{f_2'} = 100 \text{ cm.}$$

The cardinal points are plotted O to O' = 25 cm O to H = α = 100cm;

O' to H' = α' = – 50 cm;

H to F = f = – 40 cm;

H' to = F ' = f' = 40cm.

Given O to object P = – 10 cm,

we get u = H to P = – 110 cm. Now

$$\frac{1}{u} - \frac{1}{-110} = \frac{1}{40}$$

$$\Rightarrow \qquad u = \frac{440}{7} = 63 \text{ cm (H' to P')}$$

That locates P', 13 cm to the right to O'. Finally

$$m = \frac{v}{u} = \frac{400}{7} \text{x} \frac{1}{-110}$$

$$= -\frac{4}{7} = -0.57$$

One could also use Newton's method

$$x = \text{F to P} = -70 \text{ cm};$$

$$x' = \frac{(40)(-40)}{-70} = 23 \text{ cm} = \text{F' to P'}$$

$$m = -\frac{f}{x} = -\frac{-40}{-70} = -\frac{4}{7}$$

[Note that H and H' fall far from the lenses O and O' in this case. H falling far to the right makes u = – 110 cm, though O to P = = 10 cm only. Repeat the exercise with t = 35 cm to get further insight].

Example 7:

A camera objective has a positive lens of focal length 30.0 cm backed by a negative lens of focal length 6.0 cm at 25.0 cm distance. Deduce the position of the photographic plate, total length of the camera, and magnification of the image for an object 100 m is front.

Solution:

$$f_1' = 30.0 \text{ cm}, \; f_2' = -6.0 \text{ cm},$$

$$f_1' + f_2' = 24.0 \text{ cm}$$

$$t = 25.0 \text{ cm} \Rightarrow \nabla = -1.0 \text{ cm}$$

These lead to

$$f' = f_1 f_2/\nabla = 180 \text{ cm},$$

$$\alpha' = -f't/f_1' = -150 \text{ cm},$$

which locate H'(O'H' = – 150 cm) and F'(H'F' = 180 cm). Similarly,

$$f = -180 \text{ cm}, \; \alpha = -750 \text{ cm},$$

which locates H(OH = – 750 cm).

Now, given P (not shown) at 100 m to the left of O, we get

$$u = HP = -100\text{m} - (-7.5 \text{ m}) = -92.5 \text{ m}$$

$$\therefore \qquad v = \frac{uf'}{u+f'} = 1.83 \text{ m (check it)}$$

$$m = \frac{v}{u} = \frac{1.83}{-92.5} = -\frac{1}{50}$$

Since $\qquad u = H'P' = 1.83 \text{ m},$

and H'O' = – α' = 1.50 m, we get

O'P' = 32 cm

That gives length of the camera = OP' = 58 cm.

Note that the arrangement "throws", H, and H' far out towards the object. Therefore, while the focal length f ' is as large as 180 cm, the second focus F' is only 30 cm behind O'. A single lens of focal length f' = 180 cm would give about the same magnification (= – 1/50), but the camera would have to be about 2 metres long. On the other hand, a camera of length 58 cm using a single lens will need a lens of f ' = 55 cm, and its magnification would be = 55/10000 = –1/180, smaller by a factor of 3.6.

Example 8:

Determine the position of the focal points, principal points and nodal points in the case of a sphere of radius 10 cm and m = 1.5. Indicate their positions in a diagram.

Solution:

Here R_1 = 10 cm, R_2 = – 10 cm, and t = 20 cm.

$$f = \frac{-\mu R_1 R_2}{\mu(\mu-1)(R_1-R_2)-(\mu-1)^2 t}$$

$$= \frac{-1.5\times 10\text{cm}\times(-10\text{cm})}{1.5(1.5-1)(10\text{cm}+10\text{cm})-(1.5-1)^2\times 20\text{cm}} \quad f = 15 \text{ cm.}$$

$$\alpha = \frac{R_1 t}{\mu(R_1-R_2)-t(\mu-1)}$$

$$= \frac{10\text{cm}\times 20\text{cm}}{1.5[10\text{cm}-(-10\text{cm})]-20\text{cm}(1.5-1)} = +\ 10 \text{ cm}$$

$$\beta = \frac{R_2 t}{\mu(R_1-R_2)-t(\mu-1)}$$

$$= \frac{20\text{cm}(-10\text{cm})}{1.5[10\text{cm}-(-10\text{cm})]-20\text{cm}(1.5-1)} = -\ 10 \text{ cm.}$$

Therefore, the first principal point P_1 is at a distance of 10 cm to the right of the first refracting surface and the second principal point P_2 is at a distance of 10 cm to the left of the second refracting surface. As the medium on the two sides of the sphere is the same, the nodal points coincide with the principal points. Therefore, P_1, P_2, N_1, and N_2 lie at the centre of the sphere and F_1 and F_2 represent the focal points.

Example 8:

Find the focal length and the positions of cardinal points of a pano-convex lens of refractive index 1.5, the radius of curvature of the curved surface being 20 cm and thickness 1.0 cm.

Solution:

Given $R_1 = \infty$,

$R_2 = -20$ cm,

$t = 1$ cm

and $\mu = 1.5$.

Example 10:

Two thin convex lenses having focal lengths 20 cm and 5.0 cm are coaxial and separated by a distance of 10 cm. Find the equivalent focal length and power of the combination.

Solution:

The equivalent focal length of the coaxial system is given by

$$f = \frac{f_1 f_2}{f_1 + f_2 - d}$$

$$f = \frac{20cm \times 5cm}{20cm + 5cm - 10cm} = 6.67 \text{ cm}$$

The power of the combination is given by

$$P = \frac{1}{f(m)} = \frac{1}{0.0667m} = 15 \text{ D}$$

Example 11:

Two thin convex lenses of focal lengths 30 cm and 10 cm are separated by a distance of 25 cm in air. Calculate the positions of the cardinal points.

Solution:

Given that $f_1 = +30$ cm,

$f_2 = +10$ cm, and $d = 25$ cm

$$f = \frac{f_1 f_2}{f_1 + f_2 - d} = \frac{30cm \times 10cm}{30cm + 10cm - 25cm} = + 20 \text{ cm}$$

Example 12:

Two lenses of powers + 2.5 D and –5 D are kept coaxially separated by 30 cm. Find the positions of cardinal points.

Solution:

The focal length of the first lens,

$$f_1 = \frac{1}{P} = \frac{1}{2.5D} = \frac{1m}{2.5} = 40 \text{ cm}$$

The focal length of the second lens,

$$f_2 = \frac{1}{P} = -\frac{1}{5D} = -\frac{1m}{5} = -20 \text{ cm}$$

The equivalent focal length of the coaxial system is given by

$$f = \frac{f_1 f_2}{f_1 + f_2 - d}$$

$$\therefore \quad f = \frac{40cm \times (-20cm)}{40cm - 20cm - 30cm} = + 80 \text{ cm}$$

$$\alpha = +\frac{df}{d_2} = -\frac{(30cm) \times (80cm)}{-20cm} = -120 \text{ cm}$$

The first principal point P_1 is at a distance 120 cm to the left of the first lens.

$$\beta = -\frac{df}{f_1} = -\frac{-30cm \times 80cm}{40cm} = -60 \text{ cm}$$

The second principal point P_2 is at a distance of 60 cm to the left of second lens.

Example 13:

A convex lens of 10 cm focal length is placed in air from a concave lens of focal length 20 cm, at a distance of 5 cm. Find the distance between the two principal points of the combination.

Solution:

Given that $f_1 = + 10$ cm,

$f_2 = - 20$ cm and $d = 5$ cm

The equivalent focal length of the coaxial system is given by

$$f = \frac{f_1 f_2}{f_1 + f_2 - d}$$

$$\therefore \quad f = \frac{10cm \times (-20cm)}{10cm - 20cm - 5cm} = 13.3 \text{ cm}$$

$$\alpha = +\frac{df}{f_2} = \frac{(5cm) \times (13.3cm)}{-20cm} = -\ 3.32 \text{ cm}$$

The first principal point P_1 is at a distance 3.32 cm to the left of the first lens.

$$\beta = -\frac{df}{f_1} = \frac{(5cm) \times (13.3cm)}{10cm} = -\ 6.65 \text{ cm}$$

The second principal point P_2 is at a distance of 6.65 cm to the left of the second lens.

The distance between the two principal points of the combination

$$= \beta - \alpha - d = [-\ 6.65 - (-\ 3.32) - (-\ 5)] \text{ cm} = 1.67 \text{ cm}$$

Example 14:

A coaxial lens system placed in air has two lenses of focal lengths 3F and F separated by a distance 2F. Find the positions of the cardinal points.

Solution:

$$f_1 = 3F,\ f_2 = F,\ d = 2F$$

$$f = \frac{f_1 f_2}{f_1 + f_2 - d}$$

$$f = \frac{3F \times F}{3F + F - 2F} = \frac{3F \times F}{3F + F - 2F}$$

$$\alpha = +\frac{df}{f_2} = \frac{2F.(3F/2)}{F} = +\ 3F$$

Therefore, the first principal point P_1 is at a distance 3F to the right of the first lens.

$$\beta = -\frac{df}{f_1} = \frac{-2F.3F/2}{3F} = -\ F$$

The second principal point P_2 is a distance F to the left of the second lens.

The first focal point F_1 is at a distance 3F – 3F/2 = 3F/2 from the first lens and F_2 is at a distance 3F/2 – F = F/2 from the second lens. As the medium on the two sides of the lens system is the same, the nodal points N_1 and N_2 coincide with P_1 and P_2.

Example 15(a):

In a co-axial optical system the focal points are located by an experiment. Then for $x_o = -6$ *cm we find* $x'_0 = 12$ *cm and* $y'_o/y = -3/2$. *Deduce f and f'. Also predict* x'_o *and magnification m for* $x = -10$ *cm.*

Solution:

Use of Equation (8) gives

$$f = -9 \text{ cm} \qquad f' = +8 \text{ cm}$$

For the case xo = 10 cm, straight substitution gives

$$x'_o = +7.2 \text{ cm}, \; m = -0.9$$

Relationships with Principal Points as Origins. Let us now decide to measure the object and image distances from the respective principal point. If these distances are called u and v respectively, the transformation are

$$u = x_o + f \qquad y = y_o$$

$$v = x'_o + f \qquad y' = y'_o \qquad \text{...(i)}$$

Equations (i) now give

$$\frac{f}{u} + \frac{f'}{u} = 1 \qquad \text{...(ii)}$$

$$m = \frac{f}{f-u} = \frac{f'-u}{f'} \qquad \text{...(iii)}$$

Example 15(b):

Two identical thin convex lenses of focal lengths 8 cm each are coaxial and 4 cm apart. Find the equivalent focal length and the positions of the principal points.

Also, find the position of the object for which the image is formed at infinity.

Solution:

Here, $f_1 = f_2 = 8$ cm, d = 4 cm

$$f = \frac{f_1 f_2}{f_1 + f_2 - d} = \frac{8cm \times 8cm}{8cm + 8cm - 4cm} = \frac{64}{12} \text{ cm} = 5.33 \text{ cm}$$

$$\alpha = + \frac{df}{f_2} = \frac{4cm + 5.33cm}{8cm} = 2.66 \text{ cm}$$

$$\beta = - \frac{df}{f_1} = - \frac{4cm \times 5.33cm}{8cm} = - 2.66 \text{ cm}$$

The points are plotted in Fig. 5.18. For the final image to be formed at infinity

$$\frac{1}{V} - \frac{1}{U} = \frac{1}{f}$$

As V = ∞ (given), f = 5.33 cm,

we obtain U = – f = – 5.33 cm.

But U = u – a.

∴ **u** = U + α = – 5.33 cm + (+ 2.66 cm) = – **2.67 cm.**

Therefore, the object is at a distance of 2.67 cm to the left of the first lens.

Example 15(c):

Two thin convergent lenses each of 20 cm focal length are set coaxially 5 cm apart. An image of upright pole 200 m distant and 10m high is formed by the combination. Find the position of the unit and focal planes and the image. Also find the size of the image.

Solution:

Here, $f_1 = f_2 = 20$ cm, d = 5 cm

$$f = \frac{f_1 f_2}{f_1 + f_2 - d} = \frac{20cm \times 20cm}{20cm + 20cm - 5cm} = \frac{80}{7} \text{ cm} = 11.43 \text{ cm}$$

$$\alpha = + \frac{df}{f_2} = \frac{5cm \times 11.43cm}{20cm} = 2.85 \text{ cm}$$

$$\beta = - \frac{df}{f_1} = \frac{5cm \times 11.43cm}{20cm} = - 2.85 \text{ cm}$$

We have u = – 20000 cm, α = 2.85 cm.

$$U = u - \beta a = -20000 \text{ cm} - 2.85 \text{ cm} = -20002.85 \text{ cm}$$

$$\frac{1}{V} - \frac{1}{U} = \frac{1}{f}$$

$$\beta = \frac{1}{V} = \frac{1}{f} + \frac{1}{U} = \frac{1}{11.43\text{cm}} + \frac{1}{-20002.85\text{cm}}$$

or **V = 11.4 cm**

But V = u – β

$\therefore$ u = V + β 11.4 cm – 2.85 cm = **8.55 cm.**

Therefore, the first unit plane P_1 is at a distance of 2.85 cm to the right of the lens L_1 and the second unit plane P2 is at a distance of 2.85cm to the left of the lens L_2.

The first focal plane is at a distance of 11.43 cm – 2.85 cm = 8.58cm to the left of the lens L_1 and the second focal plane is at a distance of 8.55 cm to the right of the lens L_2.

The final image is formed at a distance of 8.55 cm to the right of the lens L_2.

Magnification, $m = \frac{I}{O} = \frac{V}{U}$

As O = 1000 cm, $\frac{I}{1000\text{cm}} = -\frac{11.4\text{cm}}{20002.85\text{cm}}$

$\therefore$ **I = – 0.57 cm.**

Example 16:

A thin convex and a thin concave lens each of 50 cm focal length are coaxially situated and separated by 10 cm. Find the position and nature of the final image formed of an object placed 20 cm from the convex lens (i.e., 30 cm from the concave lens).

Solution:

Here, f_1 = + 50 cm,

f_2 = – 50 cm, d = 10 cm

$$f = \frac{f_1 f_2}{f_1 + f_2 - d} = \frac{-50\text{cm} \times 50\text{cm}}{50\text{cm} - 50\text{cm} - 10\text{cm}} = +250 \text{ cm}$$

$$\alpha = +\frac{df}{f_2} = \frac{10cm \times 250cm}{-50cm} = -\ 50\ cm$$

$$\beta = +\frac{df}{f_1} = -\frac{10cm \times 250cm}{50cm} = -\ 50\ cm$$

U = 30 cm, f = + 250 cm.

$$\frac{1}{V} - \frac{1}{U} = \frac{1}{f}$$

$$\therefore \quad \frac{1}{V} = \frac{1}{f} + \frac{1}{U} = \frac{1}{30cm} + \frac{1}{250cm} = 26.8\ cm.$$

The image is formed at a distance 26.8 cm from P_2. Therefore, the image is at a distance 13.21 cm to the left of the convex lens. The image is virtual and erect.

Example 17:

An optical system consists of two thin lenses, a convex lens of focal length 20 cm and a concave lens of focal length 10 cm and separated by a distance of 8 cm. An object 1 cm in height is placed at a distance 50 cm from the convex lens. Find the position and size of the image.

Solution:

Here, f_1 = + 20 cm, f_2 = – 10 cm, d = 8 cm

$$f = \frac{f_1 f_2}{f_1 + f_2 - d} = \frac{-20cm \times 10cm}{20cm - 10cm - 8cm} = -\ 100\ cm$$

$$\alpha = +\frac{df}{f_2} = \frac{8cm \times (-100cm)}{-10cm} = +\ 80\ cm$$

The first principal point P_1 is at a distance 80 cm to the right of the convex lens.

$$\beta = -\frac{df}{f_1} = -\frac{-8cm \times (-100cm)}{20cm} = +\ 40\ cm$$

The second principal point P_2 is at a distance of 40 cm to the right of the concave lens.

For the combined system, distance of the object from P_1

U = – (50 + 80) cm = – 130 cm.

$$\frac{1}{V} - \frac{1}{U} = \frac{1}{f}$$

$$\therefore \quad \frac{1}{V} = \frac{1}{V} = -\frac{1}{100\text{cm}} - \frac{1}{130\text{cm}} = -56.5 \text{ cm}$$

The image is at a distance of 56.5 cm to the left of P_2, *i.e.*, 8.52 cm to the left of the convex lens.

$$\frac{V}{U} = \frac{V}{U}$$

$$\therefore \quad \frac{I}{1\text{cm}} = \frac{56.5\text{cm}}{130\text{cm}}$$

or $\quad$ **I = 0.435 cm.**

The image is virtual, erect and diminished.

Example 18:

Two thin converging lenses L1 and L_2 of powers 5 D and 4 D are placed coaxially 10 cm apart. Find the focal length f of the combination and positions of principal points P_1 and P_2.

Solution:

The focal length of the first lens, $f_1 = \frac{1}{P} = \frac{1}{5D} = \frac{1\text{m}}{5} = 20$ cm

The focal length of the second lens, $f_2 = \frac{1}{P} = \frac{1}{4D} = \frac{1\text{m}}{4} = 25$ cm

The equivalent focal length of the coaxial system is given by

$$f = \frac{f_1 f_2}{f_1 + f_2 - d}$$

$$\therefore \quad f = \frac{20\text{cm} \times 25\text{cm}}{20\text{cm} + 25\text{cm} - 10\text{cm}} = 14.29 \text{ cm}$$

$$\alpha = +\frac{df}{f_1} = -\frac{10\text{cm} \times (14.29\text{cm})}{25\text{cm}} = 5.7 \text{ cm}$$

The first principal point P_1 is at a distance 5.7 cm to the right of the first lens.

$$\beta = -\frac{df}{f_1} = -\frac{-10\text{cm} \times 14.29\text{cm}}{20\text{cm}}$$
$$= -7.14 \text{ cm}$$

The second principal point P_2 is at a distance of 40 cm to the left of the second lens.

Example 19:

In a co-axial system of lenses, the first and second focal point fellate 10.0 and 22.7 cm and the principal points at 17.7 and 15.0 cm respectively, all measured from a common origin. Deduce :

(i) the focal lengths of object space and image space,

(ii) the position of the image of an object placed at position –2.7 cm.

Solution:

The locations are

F(10.0), H(17.7), F'(22.7) and H'(15.0 cm).

f = distance from H to F = 10.0 – 17.7 = – 7.7 cm.

f ' = distance from H' to F ' = 22.7 – 15.0 = 7.7 cm.

x = distance from F to – 2.7 = – 2.7 – 10.0 = – 12.7 cm.

x' = ff '/x = (–7.7) × 7.7/(–12.7) = + 4.7 cm.

Since this is the distance of image from the second focus, the position of image is 22.7 + 4.7 = 27.4 cm.

Example 20:

From a given object point A on the axis of an image forming system, the path of a ray making 20^0 angle with axis was traced. The emergent ray came out with slope angle – 4°. If n = 1.62 and n' = 1.00, deduce the magnification of the image of an object placed at A.

Solution:

Lagrange – Helmholtz law gives

$$1.00 \times y' \times \tan(-4^o) = 1.62 \times y \times \tan 20^o$$

$$\therefore \qquad m = \frac{y'}{y} = \frac{1.62 \times 3640}{0.0699} = -8.4$$

Example 21:

For a co-axial image forming system the initial and final media are water (n = 1.33) and air respectively. For a given object the lateral magnefication is –2.0. Calculate :

(i) the axial magnification, and

(ii) angular magnification.

Solution:

$$n/n' = 1.33,\ m = -2.0$$

$$mx = m^2 \frac{n'}{n} = \frac{4.0}{1.33} = 3.0$$

$$\gamma = \frac{n}{mn'} = \frac{1.33}{-2.0} = -0.67.$$

EXERCISES

1. What do you understand by the power of a lens? Calculate the power of two thin lenses of focal length f_1 and f_2 separated by a distance d.
2. Calculate the focal length of a plano-convex lens for which the radius of the curved surface is 40 cm. (μ = 1.5).
3. Calculate the focal length of a thick convex lens of radii of curvature R_1 and R_2 and thickness t. Calculate the position of the principal points.
4. Define cardinal points of a system of coaxial lenses. Show that the principal planes are conjugate planes of +1 magnification.
5. Show that the focal length of a thick lens is given by

$$\frac{1}{f} = (\mu - 1)\left[\frac{1}{R_1} - \frac{1}{R_2} + \frac{(\mu - 1)}{\mu} \cdot \frac{d}{R_1 R_2}\right]$$

   Here μ is the refractive index of the material of the lens, R_1 and R_2 are the radii of curvatures of the lens surfaces and d is the lens thickness along the axis. Discuss how the converging and diverging properties of such lenses vary with thickness.
6. A thin convex lens of focal length 20 cm and a thin concave tens of focal length 10 cm are separated by a distance 8 cm. Deduce the positions of the principal and focal points of the system.
7. The magnification if the object is placed 50 cm in front of the second lens.
8. A convex lens of focal length 24 cm (μ = 1.5) is totally immersed in water (μ = 1.33). Find the focal length of the lens in water.
9. Show that the least possible distance between an object and its real image in a convex lens is four times the focal length of the lens.

10. Deduce for a thin lens an equation connecting the focal length, the radii of curvature of the surfaces and the refractive indices of the material of the lens and surrounding the medium.

11. For a thick lens with $R_1 = +10.0$ cm, $R_2 = -6.0$ cm, n = 1.50 deduce the focal length f' as the separation t changes from nearly 0 to 32 cm in steps of 8 cm.

12. Deduce the focal length and positions of cardinal points for a glass sphere (n = 1.60) of radius 2.0 cm placed in air. (Assume the rays being limited to the paraxial region).

13. Deduce the positions of cardinal points for a hemisphere of glass (n = 1.5) of radius 2.0 cm placed with plane face in contact with water (n =1.30).

14. A double convex tens of refractive index 1.50 has radii of curvature of 10.0 cm and 15.0 cm and thickness 2.0 cm. Deduce, and plot the positions of the cardinal points.

15. A convex meniscus lens of refractive index 1.50 has radii of curvature 10.0 cm and 15.0 cm and thickness 5.0 cm. Deduce the positions of the cardinal points.

16. Two convex meniscus lenses are set coaxially with their concave faces outwards and at distance 7.0 cm. The radii of curvature are 4.0 and 5.0 for the first lens and 6.0 and 8.0 cm for the second. The thickness is 2.0cm for each lens, and the material for both has n = 1.60. Deduce the focal length of the combination.

17. Describe nodal assembly method to locate cardinal points of a lens system experimentally. Explain its working with particular reference to the characteristics of nodal points.

18. A convex lens of thickness 6 cm has radii of curvature 8 cm and 10 cm. Find the focal length and position of the focal points and the principal points. (μ = 1.5).

19. Determine the positions of the focal points, principal points and nodal points in the case of sphere of radius 12 cm. (μ =1.5).

20. A concavo-convex lens has a refractive index of 1.5 and the radii of curvature of its surfaces are 15 cm and 30 cm. The concave surface is upwards and it is filled with a liquid of refractive index 1.6. Calculate the focal length of the liquid-glass combination.

21. Two convex lenses of focal lengths 20 and 5 cm are 10 cm apart.

Calculate the power of such combination.

22. Two thin convex lenses having focal lengths 6 cm and 2 cm are co-axial and separated by a distance of 4 cm. Calculate the combined focal length and the positions of the principal planes.
23. Show that for a piano-spherical lens of any thickness one principal plane always lies oh the plane face.
24. Show that for a meniscus lens which is not too thick the principal planes always fall outside the lens thickness and on the same side.
25. Show that in a combination of two coaxial lenses the distance between the two principal planes is given by r^2/Δ.
26. What are cardinal points of a thick lens or a coaxial combination of lenses? Define them, state their characteristics and show them diagrammatically for a lens system. Explain image formation with the help of these points in the case of a coaxial lens system.
27. Two thin convex lenses of focal length f_1 cm and f_2 cm are coaxial and separated by d. show that the equivalent focal length/ of the combination is given by the relation
$$f = \frac{f_1 f_2}{f_1 + f_2 - d}$$
28. A concavo-convex lens has a refractive index of 1.5 and the radii of curvature of its surfaces are 15 cm and 30 cm. The concave surface is upwards and it is filled with a liquid of refractive index 1.6. Calculate the focal length of the liquid-glass combination.
29. Derive an expression for the equivalent focal length of a system of two thin lenses separated by a finite distance when the space between them is filled with a medium of refractive index μ ($\mu > 1$).
30. Calculate the equivalent focal length of two thin co-axial lenses separated by a finite distance.
31. Design a combination of two thin lenses for which H and H' coincide with lenses L_2 and L_1 respectively, and $f' = 6$ cm.
32. Define f ' as h/δ. For the case of refraction at a single spherical surface deduce an expression for δ and hence for f '.
34. Prove that for a combination of two thin lenses of focal lengths f_1 and f_2 separated by a distance d, the focal length of the

combination is given by

$$\frac{1}{f} = \frac{1}{f_1} + \frac{1}{f_2} - \frac{d}{f_1 f_2}$$

35. What is an equivalent lens? In what respect it is called an equivalent lens?

    Derive an expression for me power of an equivalent lens corresponding to two thin lenses of known power kept coaxially in air separated by a certain distance. Also find an expression for its position from any of the two lenses.

36. In a given optical system the initial and final media have refractive indices 1.00 and 1.50 respectively. From a common origin the focal points of object space and image space are found located at 30.6 cm and 49.4 cm respectively. If an object located at –2.7 cm forms an image at 68.2 cm, deduce f./f' and the locations of H and H'.

37. What are the principal points and principal planes? Show that the principal planes are the planes of unit linear magnification.

38. Describe the construction and working of a nodal slide and show how the nodal points of a system can be located with its help.

39. Show that for a co-axial lens system, x x' = ff' where x and x' are the respective distances of the object and the image from the first and the second focal points and f and f' are the two focal lengths. What form does the expression take when the media on the two sides of the system are the same?

40. Define cardinal points of a system of co-axial lenses. Describe how you would determine experimentally the principal planes of a combination of two thin lenses separated by a distance.

41. What are the properties of cardinal points of a co-axial lens system? Plot the cardinal points of a Huygens eyepiece. How can they be determined experimentally?

42. Define cardinal points of a coaxial lens system. Show how the principal planes of such a system can be located using the theory of the method of deviation or otherwise. State the sign convention you use.

43. Derive an expression for the equivalent focal length of a system of two thin lenses separated by a finite distance when the space

between them is filled with a medium of refractive index $\mu(\mu > 1)$.

44. Determine the positions of the focal points, principal points and nodal points in the case of sphere of radius 12 cm. ($\mu = 1.5$).
45. Two thin converging lenses of focal lengths 20 cm and 40 cm are placed coaxially 20 cm apart. An object is placed at a distance of 50 cm from the first lens. Calculate the positions of the principal points, focal points and the position of the image.
46. A three dimensional object cannot be imaged in true proportions except at magnification unity. Discuss.
47. What are nodal points and nodal planes? Give their properties. Show that the nodal planes are planes of unit angular magnification.
48. Derive expressions for the equivalent focal length and the positions of principal points and focal points of a coaxial system of two thin lenses separated by a finite distance.
49. Two similar thin convex lenses of focal lengths 10 cm each are coaxial and 5 cm apart. Find the equivalent focal length and position of the principal points. Also find the position of the object for which the image is formed at infinity.
50. Two thin converging lenses each of 30 cm focal length are set coaxially 10 cm apart. An image of an upright pole 100 metres distant and 5 metre high is formed by the combination. Find the position of the unit and focal planes and the image. Also find the size of the image.

2

PARAXIAL RAYS AND SPHERICAL MIRRORS

INTRODUCTION

A ray is said to be *paraxial* if the angle θ between the ray and the principal axis (or symmetry axis) is small. Only rays inclined at less than about 10° to the principal axis are considered as paraxial rays and they lie close to the axis throughout the distance from object to image. As the angles are small, and when expressed in radians, we can set the cosines equal to unity and the sines and tangents equal to the angles. This is known as the *small angle approximation*. Accordingly, we can set

$$\cos\theta \cong 1 \text{ for } \cos\theta,$$

$$\sin\theta \cong \theta \text{ for } \sin\theta$$

and $\tan\theta \cong \theta$ for $\tan\theta$.

This is also known as *paraxial approximation*. As we are approximating sin θ to θ, paraxial approximation is also called as the *first-order theory*. A small area in the immediate surrounding of the optical axis is known as the *paraxial region*. Gauss, in 1841, was the first to give a systematic analysis of the formation of images under paraxial approximation.

The results are known as *first-order*, paraxial or Gaussian optics. If we restrict the rays to the paraxial rays, a good image is formed with monochromatic light by any optical system. Therefore, we restrict ourselves to paraxial rays in our study of mirrors and lenses.

SIGN CONVENTION

In order to specify the position of the object and i the image, we need a reference point and sign convention. There are many sign conventions in use in different books, which cause a lot of confusion in the minds of the students. We adopt the following sign convention known as *Cartesian sign convention* in our study. This is a relatively simpler convention, widely used and easy to remember.

(a) The diagrams are drawn with the incident light travelling from left to right.

(b) The distances are measured by taking the vertex as the origin.

(c) The distances measured in the direction of the incident light are considered positive while those measured in the direction opposite to the incident light are taken as negative. (All quantities measured to the right of P are positive and all those to its left are negative.)

(d) Heights measured upward and perpendicular to the principal axis are taken as positive while those measured downward are considered negative.

(e) Angles measured clockwise, beginning at the optic axis, are negative. Angles measured counter clockwise are positive.

In order to derive the formula for spherical mirrors (and lenses), we first draw a diagram and treat all the distances and angles as positive, and from the geometry of the diagram deduce an equation relating to the quantities. When the final equation is obtained, we apply the above sign convention and change the signs in front of quantities that are negative in the figure. This procedure gives us the *algebraic equation*, which can be applied to any possible case.

Conjugate Points

If a point object placed at a distance u from the mirror gives an image at a distance υ from the mirror, then an object at a distance u from the mirror will give an image at a distance u from the mirror. Thus, the position of the object and the image are mutually interchangeable. Two points, which can interchange their positions, are known as *conjugate points*. Therefore, the two points at distances u and υ are conjugate points. The centre of curvature is a self-conjugate point. Planes on which the conjugate points lie are *conjugate planes* and the distances from the vertex to these planes are conjugate distances.

SPHERICAL MIRRORS

Mirrors need not have to be always flat; they may be curved. Spherical mirrors are of the simple type among the curved mirrors. A *spherical mirror* is a segment of a spherical surface and usually has circular edges. There are two types of spherical mirrors-*concave* and *convex* mirrors. When the reflection takes place from the inner surface of the spherical segment, then the mirror is called a *concave* mirror. If light is reflected from the outer bulging surface of the spherical segment, then the mirror is called a *convex* mirror.

The laws of reflection are equally applicable to the curved mirrors. Using these laws, we can find out the nature and position of the image when the object is situated at different distances from the mirror; we can also determine the relation between the distance of the object and image and the focal length of the mirror. In order to obtain the above information, we have to first acquaint with the basic terms and definitions used in connection with the curved mirrors.

Basic Terms

The basic terms associated with spherical mirrors are as follows:

(i) The *centre of curvature*, C is the centre of the sphere of which the mirror is a small segment;

(ii) The *vertex* (*or pole*), P is the midpoint of the mirror;

(iii) The *radius of curvature*, R is the radius of the sphere of which the mirror is a small section;

(iv) The *principal axis* is the line passing through the centre of curvature C and the vertex P. It extends indefinitely in both directions.

(v) With a concave mirror, all rays parallel to the principal axis pass through a single point F after reflection. F is called the *principal focus*. The parallel rays converge to F after reflection. However, in case of a convex mirror, the reflected rays *appear* to emerge from F behind the mirror. Thus, the reflected rays *appear to diverge* from the principal focus.

(vi) *Focal plane* is a plane passing through the point F and perpendicular to the principal axis.

(vii) The *focal length*, f of the mirror is the distance, PF, between the vertex P and the principal focus F. It will be shown later that f = R/2.

(viii) The reciprocal of (he focal length is called the *power* of the mirror. Thus, P = 1/f.

(ix) The diameter MN of the circular outline of the mirror is called the *aperture* of me *mirror*. The unobstructed surface area of the mirror, having circular rim and available for reflection, represents the aperture. It is measured in terms of the angle θ, subtended by the extremities of the mirror at the centre of curvature. The aperture determines the amount of light energy that is received by the mirror. In other words, it determines the light gathering capa-bility of the mirror.

SPHERICAL MIRROR EQUATION

We now consider the image formation by a spherical mirror. Aconcave mirror with radius of curvature R facing the incident light. The *centre of curvature* of the surface is at C and the *vertex* of the mirror is at P. The line CP is the *optic axis*. Point O is an object point that lies on the optic axis and we assume that the distance from O to P is greater than R.

A ray from O, at an angle a with the axis, strikes the mirror at Q and is reflected, the incident and reflected rays making equal angles θ with the normal CQ. A second ray from O, passing through C, strikes the mirror normally and is reflected back on itself. The two reflected rays intersect at I, forming a real image of O. The image is *real* since the light energy passes through the image position. Let us now determine the location of the image point I. The basic equation governing spherical mirrors relates three lengths, namely object distance, u, image distance, υ, and the radius of curvature of the mirror, R. An exterior angle of a triangle equals the sum of the two opposite interior anglcs. Applying this theorem to the triangle OQC, we have

$$\gamma = \alpha + \theta \qquad ...(4)$$

Applying to the triangle OQI, we have

$$\beta = \alpha + 29 \qquad ...(5)$$

Eliminating angle 9 from these equations, we get

$$\alpha + \beta = 2\gamma \qquad ...(6)$$

Let h represent the height of point Q above the optic axis, and let δ represent the short distance BP.

From $\Delta^{le}OQB$, $\tan\alpha = \dfrac{h}{u-\delta}$

From $\Delta^{le}IQB$, $\tan\alpha = \dfrac{h}{u-\delta}$, and

From $\Delta^{le}CQB$, $\tan\gamma = \dfrac{h}{R-\delta}$, and

In the paraxial ray approximation, the angles α, β and γ are very small and we can replace tan α by α and so on. When α is small we can also neglect the distance x in comparison to u, υ and R.

$$\tan\alpha \cong \alpha = \frac{h}{u};$$

$$\tan\beta \cong \beta = \frac{h}{\upsilon}$$

and $\tan\gamma \cong \gamma = \dfrac{h}{\upsilon}$

Using the values of α, β and γ into the equation (6) and on simplification, we get the *spherical mirror equation*

$$\frac{1}{u} + \frac{1}{\upsilon} = \frac{2}{R} \qquad ...(7)$$

The above equation represents the *numerical relationship* between u, υ and R. We now assign the signs to these quantities as per the convention mentioned earlier. The pole P is taken as the origin and the line CP as x-axis. The rays are incident from left to right. We take the direction from left to right as the positive x-direction. The points I,C, and O are situated to the left of the origin P in the figure. Therefore, the quantities u, υ and R are negative. Putting the signs in Eq. (7), we get,

$$\frac{1}{-u} + \frac{1}{-\upsilon} = \frac{2}{-R}$$

or

$$\frac{1}{u} + \frac{1}{\upsilon} = \frac{2}{R} \qquad ...(8)$$

Eq. (8) is known as the *spherical mirror equation.* It is the algebraic relationship which incorporated the signs. Though, it is derived for a special situation, it is valid in all other situations with a spherical mirror.

Equation (8) is an approximation valid for paraxial rays only. Since the angles α, β, and γ do not appear in this equation, we conclude that, all paraxial rays passing through object point O will go through the same image point I after they are reflected at the surface. Since all such reflected light rays converge on the image point, a concave mirror is also called a *converging mirror*.

In an actual case the rays can be made more and more paraxial by putting a circular diaphragm in front of the mirror, centered about the vertex P. As a result, the image will become sharper. But, as the diaphragm *decreases the aperture* of the mirror, the image will become fainter because less total light energy will be reflected from the mirror. A compromise must often be made between image quality and image brightness.

Notes :

(i) The initial equation (7) is a *numerical relationship* between the various distances for a particular type of spherical mirror. We then adopted a sign convention for the distances, which enabled us derive an *algebraic relationship*. Such an algebraic relation (8) obtained with the sign convention adopted by us, will serve for all types of mirrors used in all possible ways. Any sign contention that is different than that adopted by us is likely to give a different algebraic relationship.

(ii) We have to bear in mind that u, υ, and R are algebraic quantities and have a sign, numerical value and a unit. We have to assign the proper sign and value for these parameters while working out the numerical problems.

(iii) For larger values of α, the reflected rays pass through different points of the axis. Eq. (7) is to be corrected and a first correction to it gives

$$\frac{1}{u} + \frac{1}{\upsilon} = \frac{2}{R} + \frac{h^2}{R}\left(\frac{1}{R} - \frac{1}{u}\right)^2$$

which shows that υ distance h and h and therefore there is no unique point image.

FOCAL POINT AND FOCAL LENGTH

If the object O is taken at a very large distance from the mirror (u = ¥), the incoming rays become almost parallel. Then, the image distance v is given by

$$\frac{1}{\infty} + \frac{1}{\upsilon} = \frac{2}{R}. \text{ Therefore, } \upsilon = \frac{R}{2}$$

The incident parallel rays converge, after reflection from the mirror, to a common point F at a distance R/2 from the vertex. The point F at which the incident parallel rays converge is called *focal point.* The distance from the vertex to the focal point is called *focal length* and is denoted by f. Focal length is defined as the image distance when the object is at infinity. Thus, $\upsilon = f$ and

$$f = \frac{R}{2} \qquad ...(1)$$

Thus, the focal length of a concave spherical mirror is half the radius of curvature. Now, let us look at the opposite situation. If the object is placed at the focal point F, the object distance u = R/2. The image distance is given by

$$\frac{2}{R} + \frac{1}{\upsilon} = \frac{2}{R}$$

or $$\frac{1}{\upsilon} = 0, \text{ therefore } \upsilon = \infty.$$

With the object at the focal point, the reflected rays are parallel to the optic axis, so the image is at infinity.

We usually express the relationship between object and image distances in terms of the focal length f:

$$\frac{1}{u} + \frac{1}{\upsilon} = \frac{1}{f} \qquad ...(2)$$

This is known as the *Gauss formula* for a spherical mirror. Since f can be determined experimentally by observing the point of convergence of rays that are parallel to the principal axis, it is not necessary to know the radius of R in order to apply Eqn. (2).

Mathematically we can interchange u and υ and still satisfy the mirror equation. It illustrates the principle of *optical reversibility*, which states that if any ray is reversed, it will retrace its path through an optical system.

EXTENDED OBJECT

Now, suppose we have an object with finite size, represented by the arrow AB, perpendicular to the optic axis. A real and inverted image A'B' is formed between C and F.

Δ^{les} ABC and A'B'C are similar and hence

$$\frac{AB}{A'B'} = \frac{CD}{CB'} \quad ...(1)$$

Also, Δ^{les} ABP and A'B'P are similar.

Hence $$\frac{AB}{A'B'} = \frac{PB}{PB'} \quad ...(2)$$

From equations (1) and (2), we get

$$\frac{CB}{CB'} = \frac{PB}{PB'}$$

or $$\frac{PB - PC}{PC - PB'} = \frac{PB}{PB'}$$

$\therefore$ $$\frac{u - R}{R - \upsilon} = \frac{u}{\upsilon}$$

$\therefore$ $$uR + \upsilon R = 2u\upsilon$$

Dividing throughout by $u\upsilon R$, we get

$$\frac{1}{u} + \frac{1}{\upsilon} = \frac{2}{R} \quad ...(3)$$

Using the sign convention, u = –ve, R = –ve and υ = –ve into the above equation, we get

$$\frac{1}{-u} + \frac{1}{-\upsilon} = \frac{2}{-R}$$

$\therefore$ $$\frac{1}{u} + \frac{1}{\upsilon} = \frac{2}{R} \quad ...(4)$$

Using the relation into (4), we obtain

$$\frac{1}{u} + \frac{1}{\upsilon} = \frac{2}{f} \quad ...(5)$$

This relation is the *Gauss' formula.*

LATERAL MAGNIFICATION

The ratio of the transverse dimensions of the final image formed by an optical system to the corresponding dimension of the object is defined as the *transverse* or *lateral magnification.* It is noted that the object and image have different sizes and that they have opposite orientations. Because triangles ABP and A'B'P are similar, we have

$$\frac{h}{u} = \frac{h'}{\upsilon}$$

$$\therefore \qquad m = \frac{h'}{h} = -\frac{\upsilon}{u} \qquad \text{...(1)}$$

Note : The term magnification is generally used in a wider sense. Magnification does not mean that the image is enlarged. The image formed by an optical system (mirror, lens etc.) may be larger than, smaller than, or of the same size as the object.

Convex Mirror

The formation of image of a point object by a convex mirror. A ray from O, incident along the axis, strikes the mirror normally at P and is reflected back on itself. A second ray from O, at an angle a with the axis, strikes the mirror at Q and is reflected, the incident and reflected rays making equal angles 6 with the normal CQ. The two reflected rays, when extended backward, intersect at I, forming a virtual image of O. The image is virtual since the light energy does not pass through the image position. Following the same procedure that we used for a concave mirror, it is easy to show that for a convex mirror.

$$\frac{1}{u} + \frac{1}{\upsilon} = \frac{2}{R}.$$

Graphical Method

We can determine the properties of the image by a simple *graphical method.* This method consists of finding the point of intersection of a few particular rays that diverge from a point of the object and are reflected by the mirror. Then *all* rays from this point that strike the mirror will intersect at the same point. For this construction we always choose an object point that is not on the optic axis. Out of all possible light rays that can be drawn from a point on an object, there are four rays, which are useful in locating the corresponding image point. These are called *principal rays* or *easy rays.*

(a) A *ray parallel to the principal* axis, after reflection, passes through the focus F in case of concave mirror; or appears as though it came from the focal point F in case of convex mirror.

(b) A *ray heading towards or away from the centre of curvature* C is reflected back along its original path. Because the radius and tangent at the point of incidence are at 90° to one another, any ray incident normally on a mirror is reflected along the same path.)

(c) A *ray incident at vertex* P at an angle i is reflected at an angle i.

(d) A *ray through the focal point* F is reflected parallel to the axis.

Normally, any of the two rays mentioned above are sufficient to locate the image of a linear object. However, a third ray is often used as a safeguard against errors.

EFFECT OF ALTERING THE OBJECT DISTANCE

We will now consider the location of an image graphically when the object lies at various distances from the mirror. In each case we shall find out the position, nature and relative size of the image.

(i) Object at Infinity

Since the object is at infinity, the incident rays become parallel. Two such rays are shown in the figure, one along the principal axis and the other parallel to it. The ray parallel to optic axis is incident on the mirror at N and the reflected ray passes through F. The ray along the principal axis is reflected at P and retraces its path. The two reflected rays intersect at F and the image is formed at point F. A concave mirror may form a real or virtual image of an object depending upon the distance of the object from the mirror.

The reflected rays actually pass through the image points in case of a *real-image* and the image can be projected onto a screen. Note that as the object moves toward the mirror, the image recedes away from the mirror and finally appears behind the mirror. As the object approaches the focus of the mirror, the image recedes further from the mirror and when the object is at focus, the image is at infinity. When the object is within the focal length of the mirror, the image is virtual, erect, magnified and is formed behind the mirror. When the object is at C, the image formed is real, inverted and of the same size as the object. Further the image is also situated at C.

In a convex mirror, no matter where the object is situated in front of the mirror, the image is formed behind the mirror. The image is virtual, erect, diminished and is always formed between the vertex P and the focus F. As the object moves toward the mirror, the image formed behind the mirror shifts toward the vertex. When the object is at a distance equal to the focal length of the mirror, the image is formed exactly at the midpoint between the vertex and the focus. The convex mirror has a wide field of view and hence is used as a driving mirror.

Caution

In actual practice the paraxial condition is difficult to ensure. The presence of large angle rays, falling on a spherical mirror of wide aperture, will not converge to or diverge from a single point. Hence, the image of an extended object appears blurred. The lack of image sharpness caused by the spherical mirror surface is called the *spherical aberration.* The rays reflected by the mirror intersect along a conical surface and produce a bright curves, called the *reflection caustic curve.* Such caustic curves are seen under appropriate illumination condition on the surface of milk or tea in a cup.

Spherical aberration cannot be eliminated completely, but by proper design of the surface it can be suppressed for certain positions, called *anastigmatic.* Anastigmatic positions can be modified by changing the shape of the surface. An *elliptical mirror* is anastigmatic for an object placed at one focus of the ellipse whose image falls exactly at the other focus. The two foci thus are conjugate. Such mirrors are used for optical pumping of ruby laser. Similarly, a *parabolic mirror* produces no aberration for rays that are parallel to the principal axis. All these rays will pass through the focus of the parabolic mirror. Such mirrors are used in astronomical telescopes, search lights, and microwave antennas.

REFRACTION THROUGH A GLASS SLAB

A light ray that travels through a parallel-sided transparent slab (*e.g.*, plate glass window) is refracted at both faces. The two refractions at the parallel surfaces result in a 'sideways' displacement, but do not change the direction of the ray. To show that there is no change in direction, we can apply Snell's law at both faces.

At the first face, the Snell's law takes the form

$$u_i \sin \theta_1 = \mu_2 \sin \theta_2 \qquad ...(1)$$

Because the faces are parallel, the angle of incidence for the second face equals θ_2 and according to Snell's law

$$\mu_2 \sin \theta_2 = \mu_1 \sin \theta_3 \qquad ...(2)$$

By comparing Eqns. (1) and (2), we that $\theta_1 = \theta_3$. When a ray of light passes through a parallel-sided slab, the emergent ray is parallel to the incident ray.

The net effect of the two refractions is a parallel displacement of the ray. The distance x between the incident and emergent ray is given by

$$x = \frac{t.\sin(\theta_1 - \theta_2)}{\cos\theta_2} \quad ...(3)$$

where t is the thickness of the glass slab.

It can also be shown that $x = t\theta\left(\frac{\mu - 1}{\mu}\right)$ ...(4)

for small angles θ and θ measured in radians.

For an ordinary windowpane the displacement is generally not more than a few millimeters.

REFRACTION THROUGH A COMPOUND SLAB

For a single parallel-sided slab, the emergent beam will be parallel to the incident beam. If r is the angle of refraction, the Snell's gives

$$\frac{\sin i}{\sin r} = {}_1\mu_2 \text{ and } \frac{\sin r}{\sin i} = {}_2\mu_1 \quad ...(1)$$

$${}_1\mu_2 \cdot {}_2\mu_1 = 1 \quad ...(2)$$

For a number of parallel-sided slabs, placed one upon the other, the Snell's law gives

$$\frac{\sin i}{\sin r_1} = {}_1\mu_2, \quad \frac{\sin r_1}{\sin r_2} = {}_2\mu_3,$$

and $$\frac{\sin r_2}{\sin i} = {}_3\mu_1$$

$$\therefore \quad {}_1\mu_2 \cdot {}_2\mu_3 \cdot {}_3\mu_1 = 1$$

$$\therefore \quad {}_2\mu_3 = \frac{1}{{}_1\mu_2 \cdot {}_3\mu_1} = \frac{{}_1\mu_3}{{}_1\mu_2} \quad ...(3)$$

Note : The visibility of a transparent medium is due to the difference in its refractive index from that of the surrounding medium. A piece of transparent glass placed in water is visible because of the difference between their refractive indexes.

Apparet Depth

Let, an object lie below a refracting medium of refractive index μ_2 and is seen almost normally in a medium of refractive index μ_1.

According to Snell's law, we have

$\Rightarrow$ $$\mu_2 \sin i = \mu_1 \sin r$$

or $$\mu_2 \frac{AB}{OB} = \mu_1 \frac{AB}{IB}$$

Let OB = u and IB = υ.

Then $$\mu_2 \upsilon = \mu_1 u$$

$\Rightarrow$ $$\upsilon \frac{\mu_1}{\mu_2} \mu$$

If $\mu_2 > \mu_1$, $\upsilon < u$, and the image I of an object O appears to be raised through a distance OI. This distance is given by

$$\Rightarrow \quad OI = t\left(1 - \frac{\mu_1}{\mu_2}\right), \qquad ...(4)$$

where t is the thickness of the medium,

In case of viewing other than vertically, the position of the image changes with the angle, which the line of vision makes with the surface of separation of the two media. As the angle increases, the apparent depth decreases. Also, the object appears curved when viewed obliquely through the refracting medium.

INTERNAL REFLECTION

Light is partially reflected and partially transmitted at an interface between two materials with different indexes of refraction. Under certain circumstances, however, all of the light can be reflected back from the interface, with none of it being transmitted, even though the second material is transparent. Several rays radiating from a point source P in material having a refractive index μ_2. The rays strike the surface of a second material having refractive index μ_1 ($\mu_2 > \mu_1$). From Snell's law of refraction

$$\sin \theta_2 = \frac{\mu_1}{\mu_2} \sin \theta_1 \qquad ...(1)$$

Because μ_2/μ_1 is greater than unity, $\sin \theta_2 > \sin \theta_1$. Therefore, the ray is bent away from normal. Thus, there must be some value of θ_1 less than 90° for which Snell's law gives $\sin \theta_2 = 1$ and $\theta_2 = 90°$. Ray 3 in the diagram emerges just grazing the surface at an angle of refraction Of 90°.

The angle of incidence, for which the refracted ray emerges tangent to the surface, is called the *critical angle*, denoted by θ_{crit}. If the angle of incidence is *greater than* the critical angle, the sine of the angle of refraction, as computed by Snell's law, would have to be greater than unity, which is impossible. Beyond the critical angle, the ray *cannot pass* into the upper material; it is trapped in the lower material and is completely reflected at the boundary surface. This is called *total internal reflection.*

We can find the critical angle for two given materials by setting $\theta_2 = 90°$ in Snell's law. We then have

$$\sin \theta_{crit} = \frac{\mu_1}{\mu_2} \qquad ...(2)$$

Light propagating within the glass ($\mu \cong 1.52$) will be totally reflected if it strikes the glass-air surface at an angle of about 41° or more.

When a beam of light enters at one end of a transparent rod, the light can be totally reflected internally, if the index of refraction of the rod is greater than that of the surrounding material. The light is trapped with in the rod even if the rod is curved. Such a rod is called a light pipe.

APLANATIC POINTS

A spherical lens, which is free from the defects of spherical aberration and coma is called an *aplanatic lens.* A pair of conjugate points free from spherical aberration and coma is called aplanatic points. The property of an aplantic lens. Let O be the center of curvature of the lens of refractive index μ and radius of curvature R. P is a point on the axis of the lens such that PO = R/μ. It can be shown that all rays passing through the point P appear to diverge through the point Q irrespective of the slope angle made by the incident rays. PA is the incident ray and AC is the refracted ray. The ray AC appears to diverge from the point Q which is the image of P. Let i and r be the angles of incidence and refraction and a and P the slope angles made by the incident and refracted rays.

Then,
$$\frac{\sin i}{\sin r} = \frac{1}{\mu} \qquad ...(1)$$

In the Δ APO,
$$\frac{\sin i}{\sin \alpha} = \frac{PO}{R} = \frac{R}{\mu R} = \frac{1}{\mu} \qquad ...(2)$$

From (1) and (2)

$$\frac{\sin i}{\sin r} = \frac{\sin i}{\sin \alpha}$$

$$\therefore \qquad \angle\alpha = \angle r \qquad ...(3)$$

In the ΔAPQ,

$$\angle\alpha = \beta + (r - i) \qquad ...(4)$$

Substituting $\angle\alpha = \angle r$ in equation (10), we get

$$r = \beta + r - i \text{ or } i = \beta$$

In the ΔAOQ,

$$\frac{\sin r}{\sin \beta} = \frac{\sin r}{\sin i} = \mu = \frac{OQ}{OA} = \frac{OQ}{R}$$

$$\therefore \qquad OQ = \mu.R.$$

Thus, if the distance of the object point P is R/μ from the centre of curvature, then distance of the image point Q is μR irrespective of the slope angles α and β. The object and image distances of the conjugate points that satisfy the Abbe condition (6)

$$BP = R + \frac{R}{\mu}, \text{ and } BQ = R + \mu R. \qquad ...(5)$$

The object and image points which are at distances given by eqn. (5) from the pole of the spherical surface are free from spherical aberration as well as coma. Such points are known as *aplanatic points*.

An aplanatic lens is mostly used as the front lens of a high power microscope objective called the oil immersion objective. As it is not possible to place an object inside a solid spherical lens, the lens is ground a little and the object to be examined is embedded in between a drop of oil and the lens surface. The oil chosen is such that it has the same refractive index as that of the lens.

Aplanatic Surfaces

Consider an image forming system in which a *very wide cone* of rays from a point P comes to sharp focus at another point P'. The points concerned are called aplanatic points and the system concerned is called aplanatic system. For a given lens or reflector this condition of zero spherical aberration for a wide cone of rays applies *for a particular pair of conjugate points* only, not for all the conjugate pairs of points.

Let P and P' be a pair of points and consider the locus of a point A such that

$$PA + AP' = \text{Constant} = K.$$

This locus of A is an ellipse. One of the properties of ellipse is that

PA and AP' make equal angles with the normal at A Therefore, if the ellipse is rotated about the axis PP' and the surface so obtained (ellipsoid) is silvered, we get a mirror such that *all* the rays from P, whatever their angle, convex sharply at P' and *vice versa.* The ellipsoidal surface is called an a plane mirror, and the pair of points P, P', which are its foci, are called aplanatic points.

A parabolic mirror is a special case of the above one, where aplanatic point is its focus, the other lies at infinity. Such mirrors are used in motor headlights, lighthouses, etc.

We show a spherical medium with centre O, radius r and refractive index n, surrounded by air. Consider an object point Pat distance OP = r/n. Consider *any* ray PA from P. It is refracted along AB. Produced meet the line OP at P'. We propose to calculate OP'.

In ΔOPA,

$$\frac{\sin \theta}{\sin \alpha} = \frac{OA}{OP} = \frac{r}{r/n} = n$$

But, by Snell's law sin β/sin α = n. Hence, H follows that β = θ.

Then in ΔAP'P

$$\phi = 0 - (\beta - \alpha) = \beta - (\beta - \alpha) = \alpha$$

This gives

$$\frac{OP'}{OA} = \frac{\sin \beta}{\sin \phi} = \frac{\sin \beta}{\sin \alpha} = n \qquad ...(6)$$

We note that distance OP' is independent of the inclination (θ) of the incident ray PA from the axis PO. No approximation is involved. So all rays from P, *whatever their slope*, will, after refraction, appear to diverge from P'. These conjugate points P, P' are called aplanatic points of the sphere.

If we draw two spheres around O with radii OP and OP', then we shall find that for every object point on the first sphere there will be an image point on the second sphere free from spherical aberration. These two surfaces are called aplanatic surfaces. (They are not plane surfaces). The linear magnification is rn r/n = n^2, and surface magnification is n^4.

Practical use of aplamtic points of a sphere is made through the use of (i) an oil immersion lens, and (ii) a 'meniscus' lens; *Oil Immersion Lens :* This is used in microscope objectives. The objective is a hemispherical lens of radius r and refractive index n. The object is placed on a platform at depth r/n, and a drop of oil of refractive index n fills the space between the platform and the lens. As discussed above, there is no spherical aberration for object point at P.

One can see that the semi-angle θ of the cone of incident rays from P entering the lens is given by tan θ = n, while the semi-angle θ' of the arrangement cone of rays is given by tan θ' =1/n, so that the cone of rays converges by a factor tan θ/tan θ = n^2.

Aplamtic Meniscus Lens :The object point P to be placed "inside" the sphere. Meniscus lens is a way of achieving this. Face II has radius r and its centre is at O. Face I has its centre at P, with distance OP = r/n. All rays from P are thus normal to Face I and suffer refraction only at Face II. They form image at P'. Thus P, P' form an aplanatic pair of points.

ELIMINATION OF SPHERICAL ABERRATION

We now briefly record the different methods of reducing or eliminating spherical aberration :

(i) *Reducing the Lens Aperture* : Reducing the aperture h reduces SA considerably. But it also reduces intensity of the image and resolving power of the instrument.

(ii) *"Bending" of a Lens* : For given n and given f of a lens the geometrical quantity $\left(\frac{1}{R_1} - \frac{1}{R_2}\right)$ is fixed. But individual radii R_1 and R_2 can be chosen variously. Spherical aberration is a function of the shape factor $R_1/R_2 = \beta$. We can therefore choose β in such a way that the aberration is minimum. This process of choosing β is called *"bending" of the lens*. When the bending exactly suits the minimum aberration condition, the lens is said to be a *"crossed lens."*

(iii) *Use of Separated Doublet* : When a separated doublet is used, the separation between the lenses for minimum spherical aberration is given by,

$$t = f'_1 - f'_2$$

Additional advantage conies if each lens is itself "bent" to give minimum spherical aberration. This condition is used, for example, in Huygens' eyepiece.

(iv) *Contact Doublet of Convex and Concave Lenses of Suitable*

Shape Factors : If K represents [ϕ < (n, β)], so that SA = Kh^2/f, then a contact double t has zero SA when

$$\frac{K_2}{K_1} = -\left(\frac{f_2'}{f_1'}\right)^3 \qquad ...(13)$$

This condition can be fulfilled together with the condition of achromatism because the shape factor is a free parameter for adjusting values. But K depends also on u, υ hence, the condition holds for a given pair of conjugate points only.

(v) *Use of Aplanatic Points :* Property of aplamtic points is useful when the object can always be brought to a particular position, *e.g.,* in a microscope objective. In this case *spherical aberration is totally absent*, but this holds only for one pair of conjugate points on the axis. For other pairs of conjugate points spherical aberration exists, but is small (coma is more serious here).

(vi) *Use of Specially Corrected Surfaces :* Mirror or lens surfaces may be specially "corrected" in such a way that rays from all zones come to a common focus. The surface is then not spherical and is called "a spherical." Such arrangements are used in all expensive telescopes and cameras using large aperture. However, the correction will be applicable for one pair of conjugate points only—hence for each special task special correction has to be worked out and executed.

An alternative is to use *'corrector plates.'* Fig 9.20 shows the Schmidt corrector plate P used widely with large aperture reflector telescopes. For a spherical reflector, the rays falling on outer parts (larger A) come to focus earlier than those falling on the central part of the mirror. The Schmidt plate effectively compensates this by acting as a concave lens for rays at larger h and as a convex lens for rays at smaller h. Its positioning is at the centre of curvature of the reflector R to ensure that the effect is independent of the inclination of the chief ray.

REFLECTING PRISMS

Prisms perform different roles in optics. We broadly classify them into two major groups as *reflecting prisms* and *dispersing prisms*. We first study the reflecting prisms. Reflecting prisms make use of *total internal reflection* and are employed for the purpose of changing either the direction of light or the orientation of the image or both. Because the critical angle for glass is slightly less than 45°, it is possible to use a prism with angles of 45°–45°–90° as a totally reflecting surface. As reflectors, totally reflecting prisms have some advantages over metallic surfaces such as coated glass mirrors. While no metallic surface reflects 100% of the light incident on it, light can be totally reflected by a prism. The reflecting properties of a prism have the additional advantage of being permanent and unaffected by tarnishing.

A light beam is incident on the prism in such a way that at least one internal reflection takes place within the prism and emerges without dispersion. Some of the commonly used reflecting prisms are discussed here.

The *right-angled prism* is a 45°–45°–90° prism. The light is reflected at the hypotenuse and rays are deviated normal to the incident face by 90°. It is easy to see that the top and bottom of the image have been interchanged but the right and left sides have not. It means that the top surface acts like a plane mirror.

The *Porro prism* is physically the same as the right angle prism but is used in a different orientation. Light enters and leaves at right angles to the hypotenuse and is totally reflected at each of the shorter faces. The total change of direction of the rays is 180°. Thus, if enters right-handed, it leaves right-handed. Binoculars use combinations of two Porro prisms.

The *Dove prism* is a truncated version of the right-angle prism, used almost exclusively in collimated light. As the prism is rotated about the line of sight, the image rotates through twice the angle. The *penta prism* deflects the light beam through a constant 90°, without affecting the orientation of the image. It is essential that two of its surfaces must be silvered. These prisms are often used as end reflectors in range finders.

The *comer-cube* prism is obtained if one cuts a cube so that the piece removed has three mutually perpendicular faces. It has the property of being retrodirective; that is, it will reflect all incoming rays back along their original directions.

REDUCTING SPHERICAL ABERRATION

Spherical aberration produced by lenses is minimized or eliminated by the following methods:

1. Spherical aberration can be minimized by using stops, which reduce the effective lens aperture. The *stop* used can be such as

to permit either the axial rays of light or the marginal rays of light. However, as the amount of light passing through the lens is reduced, correspondingly the image appears less bright.

2. The longitudinal spherical aberration produced by a thin lens for a parallel incident is given by,

$$x = \frac{\rho_2}{f_2}\left[\frac{k^2\mu^2 + k\left(\mu + 2\mu^2 + 2\mu^3\right) + \mu^3 - 2\mu^2 - 2}{2\mu(\mu-1)^2\,(1-k)^2}\right] \qquad ...(2)$$

where, x is the longitudinal spherical aberration, ρ is the radius of the lens aperture and f_2 is the second principal focal length.

$$k = \frac{R_1}{R_2}$$

where R_1 and R_2 are the radii of curvature. For given values of μ, f_2 and ρ, the condition for minimum spherical aberration is

$$\frac{dx}{dk} = 0.$$

Differentiating eqn. (2) and equating the result to zero, we get

$$k = \frac{R_1}{R_2} = \frac{\mu(2\mu-1)-4}{\mu(2\mu+1)} \qquad ...(3)$$

From equation (3), for a lens whose material has a refractive index $\mu = 1.5$, $k = -1/6$. Thus, the lens, which produces minimum spherical aberration, is biconcave and the radius of curvature of the surface facing the incident light is one-sixth the radius of curvature of the other face.

In general, the more curved surface of the lens should face the incident or emergent beam of light which ever is more parallel to the axis. A lens whose $\left(\frac{R_1}{R_2} = -\frac{1}{6}\right)$, is called a cross lens. The process in which the shape of the lens is changed without changing the focal length of the lens is called bending of the lens for minimum spherical aberration. A crossed lens is shown in Fig. 9.4.

It is clear from the figure that the deviation produced by the two surfaces is the same and the axial and marginal rays of light come to focus with minimum of spherical aberration. However, it should be noted that the spherical aberration cannot be completely eliminated in a lens with spherical surfaces. For a lens of refractive index 1.5, focal length 100 cm and radius of the lens aperture 10 cm, the longitudinal spherical aberration is 1.07 cm for k = –1/6. For the same values of focal length and radius of the lens aperture, if the values of p and k are 2 and +1/5, the longitudinal spherical aberration reduces to 0.44cm.

3. Plano-convex lenses are used in optical instruments so as to reduce the spherical aberration. When the curved surface of the lens faces the incident or emergent light whichever is more parallel to the axis, the spherical aberration is minimum. The spherical aberration in a crossed lens $\frac{R_1}{R_2} = -\frac{1}{6}$ is only 8% less than that of a plano-convex lens having the same focal length and radius of the lens aperture. This is the reason why plano-convex lenses are generally used in place of crossed lenses without increasing the spherical aberration appreciably. Fig. 9.5 represents the variation of longitudinal spherical aberration with the radius of the lens aperture for lenses of the same focal length and refractive index.

The spherical aberration will, however, be very large if the plane surface faces the incident light. The spherical aberration is a result of larger deviation of the marginal rays than the paraxial rays. If the deviation of the marginal rays of light is made minimum, the focus f_n for a parallel incident beam will shift towards f_p, the focus for the paraxial rays of light and the spherical aberration will be minimum.

As the deviation is a minimum in a prism, when the angles of incidence and emergence are equal, similarly in a lens also, spherical aberration can be minimized if the total deviation produced by a lens is equally shared by the two surfaces. In a plano-convex lens, when the plane surface faces the parallel beam of light, the deviation is produced only at the curved surface and hence the longitudinal spherical aberration (Fig. 9.6) is more than when the curved surface faces the incident light (Fig. 9.7). In the latter case the spherical aberration is less than the former, because the total deviation in the second case is divided between the two surfaces. Thus, the spherical aberration produced by a single lens can be minimized by choosing proper radii of curvature. The shape factor q of lens is given by

$$q = \frac{R_1 + R_2}{R_2 - R_1} \quad ...(4)$$

The lenses of the same values of f, μ and p but of different shape factors calculated according to equation (4). The longitudi-nal spherical aberration is represented along the y-axis and the shape factor θ along the x-axis.

Spherical aberration for a double convex lens (shape factor 0.5) is a minimum when the surface of smaller radius of curvature faces the incident parallel light. The spherical aberration for a plano-convex lens (shape factor +1.0) when the curved surface faces the incident light is only slightly more than the double convex lens. Hence, piano- convex lenses are preferred.

4. Spherical aberration can also be made minimum by using two plano-convex lenses separated by a distance equal to the difference in their focal length. In this arrangement, the two lenses equally share the total deviation and the spherical aberration is minimum. In Fig. 9.9, two plano-convex lenses of focal lengths f_1 and f_2 are separated by a distance d.

   Let δ be the angle of deviation produced by each lens (Fig. 9.9).

$$\angle BF_1G = \delta, \angle F_1BG = \delta$$

and from $\Delta^{le}BGF_1$, $BG = GF_1$

or $$O_2G = GF, \text{ (approximately)}$$

$$\therefore \quad O_2\,G = 1/2(O_2F_1) = 1/2\,(f_1 - d)$$

For the second lens, F_1 is the virtual object and G is the real image. Substituting these values of object and image distances in the formula

$$\frac{1}{\upsilon} - \frac{1}{u} = \frac{1}{f_2} \qquad (u, \upsilon\ f_2 \text{ are all +ve})$$

$$\frac{1}{O_2G} - \frac{1}{O_2F_1} = \frac{1}{f_2}$$

$$\frac{2}{(f_1 - d)} - \frac{1}{(f_1 - d)} = \frac{1}{f_2}$$

$$\frac{1}{(f_1 - d)} = \frac{1}{f_1}$$

$$f_2 = f_1 - d \text{ or } d = f_1 - f_2 \qquad ...(5)$$

Thus, *the condition for minimum spherical aberration is that the distance between the two lenses is equal to the difference in their focal lengths.*

5. Spherical aberration for a convex lens is +ve and that for a concave lens is –ve. By a suitable combination of convex and concave lenses spherical aberration can be made minimum.

CONSTANT OF THE DISPERSING PRISM

Constant deviation dispersing prisms are widely used in spectroscopy. It may be regarded as made up of two 30°–60°–30° prisms and one 45°–45°–90° prism. Suppose that in the position shown a mono-chromatic ray of wavelength A, traverses the component prism DAE symmetrically, thereafter to be reflected at 45° from the face AB. The ray will then traverse prism CDB symmetrically having experienced a total deviation of 90°.

The ray can be thought of as having passed through an ordinary 60° prism at minimum deviation. All other wave-lengths present in the beam will emerge at other angles. If the prism is rotated slightly about an axis normal to the paper, the incoming beam will have a new incident angle. Now, a different wavelength component undergoes a minimum deviation, which is again 90°- hence the name constant deviation. With a prism of this sort, one can conveniently set up the light source and viewing system at a fixed angle, *i.e.* 90° and then simply rotate the prism to look at a particular wavelength.

Dispersion

Ordinary white light is a mixture of waves with wavelengths extending throughout the visible spectrum. The speed of light in vacuum is the same for all wavelengths, but the speed in a material substance is different for different wavelengths. Therefore, the refractive index changes as a function of wavelength. The angle of refraction also depends on the wavelength. The dependence of wave speed and index of refraction on wavelength is called *dispersion.* A graph of the index of refraction, μ, versus wavelength, λ, is called a dispersion curve.

If a piece of glass is in the form of a plate with parallel sides, the rays which emerge are parallel, the different colours (wavelengths) are superposed again and no dispersion is observed except at the very edges of the image. The effect is not normally noticeable. But if the light passes through a prism, the different wavelengths in white light are refracted through different angles; and the emerging rays are not parallel for the different colours and the dispersion is clearly noticeable.

The light spreads out into a fan-shaped beam, showing a band of colours. The deviation (change of direction) produced by the prism increases with increasing frequency or decreasing wavelength. Violet light is deviated most, and red is deviated least; other colours are in intermediate positions. The light is said to be *dispersed* into a *spectrum.*

The effect is known as *chromatic dispersion.* A spectacular example of the dispersion of light is the rainbow.

The amount of dispersion depends on the difference between the refractive indexes for violet and for red light. The angle subtended by the directions of the two colours is called the *angular dispersion.*

Dispersive Prisms

If the light contains more than one wavelength, a refracting prism will disperse the light. Prisms that separate light according to wavelengths are known as *dispersive prisms.* The majority of dispersive prisms are right prisms on triangular bases. Such a prism consists of two refracting surfaces inclined at some angle to one another. The two refracting faces are labelled ABED and ACFD. The angle between them is called the *refracting angle* A, or simply the *angle of the prism* and AD is called the refracting edge. The optics of the prism can be understood by considering a triangular section cut parallel to ABC and DEF. In general i_1 is not equal to i_2 δ is the angle of deviation : it is the angle that the emergent ray has to be rotated through to bring it into coincidence with the incident ray. The deviation of a ray, δ, depends on the angle of incidence and it varies as i_1 increases from 0° to 90°. It has a minimum, δ_m, at one value of i_1. When this occurs the ray travels through the prism symmetrically, that is $i_1 = i_2 = i$ is drawn when minimum deviation occurs. NR and MR are normals at the point of incidence P and the point of emergence Q respectively. The internal angles of quadrilateral APRQ add to 360°.

$$\angle A + \angle PRQ + 180° = 360°$$

$$\therefore \quad \angle PRQ = 180° - \angle ZA$$

From the triangle PRQ, $\angle PRQ = 180° - 2r \quad (r_1 = r_2 = r, \text{ say})$

$$\therefore \quad A = 2r$$

From the triangle PTQ, $\delta_m = 2(i - r) = 2i - A$

$$i = (A + \delta_m)/2$$

Now Snell's law gives $\sin i = \mu \sin r$

Hence, we obtain $$\mu = \frac{\sin\left[(A + \delta_m)/2\right]}{\sin A/2} \qquad ...(1)$$

This is the prism equation, which holds only for light deflected at the angle of minimum deviation. At the angle of minimum deviation the light passes through the prism symmetrically and the angles of incidence and emergence are equal.

Dispersive prisms are mainly used in spectrometers to separate closely adjacent lines. Prisms made out of glass are used in the visible region. Outside the visible region, glass shows absorption. Therefore, in the UV region prisms made out of quartz or fluorite are used and in the IR region, they are made of rock salt (NaCI, KCI) or sapphire (Al_2O_3).

REFRACTION AT SPHERICAL SURFACES

Images can be formed by reflection as well as by refraction. We study here the refraction at a spherical surface, that is, at a spherical interface between two transparent materials having different indices of refraction. What happens when a real object is brought successively closer to a convex refracting surface.

When the object distance is large, the rays passing through the refracting surface come together and form a real image. If the object is moved closer to the surface to the position F_1 the emergent light is parallel and the image forms at infinity. If the object is brought even closer, the rays emergent on the right hand side of the surface do not come together. Instead, the light appears to come from a point I, lying to the left side of the surface and the image is a virtual image. Now we derive the relationships between the various distances in two different situations.

(a) Refraction at a Convex Surface Forms a Real Image

In convex surface XY with radius R forms an interface between two materials with different indexes of refraction μ_1 and μ_2. Let C be the centre of curvature and P the pole of the spherical surface. Consider a point object O lying on the principal axis in the rarer medium of refractive index μ_1 at a distance u from the vertex in front of the convex surface; the distance u being greater than the radius of curvature R of the surface.

Ray OP travelling along the principal axis strikes the vertex P and passes into the second medium without deviation. Ray OA, making an angle α with the axis, is incident on the surface at an angle θ_1 with the normal and is refracted along AI at an angle θ_2. These rays intersect at I at a distance v to the right of the vertex. The case $\mu_1 < \mu_2$. We use the theorem that an exterior angle of a triangle equals the sum of the two opposite interior angles. Applying this to the triangles OAC and ACI gives

$$\theta_1 = \alpha + \gamma, \qquad \gamma = \beta + \theta_2, \qquad ...(1)$$

From the law of refraction,

$$\mu_1 \sin \theta_1 = \mu_2 \sin \theta_2$$

Also the tangents of α, β, and γ are

$$\tan \alpha = \frac{h}{u+\delta}$$

$$\tan \beta = \frac{h}{\upsilon-\delta},$$

$$\tan \gamma = \frac{h}{R-\delta}. \qquad ...(2)$$

For paraxial rays, θ_1 and θ_2 are both small in comparison to a radian, and we may approximate both the sine and tangent of either of these angles by the angle itself. The law of refraction then gives

$$\mu_1\theta_1 = \mu_2\theta 2 \qquad ...(3)$$

Combining this with the first of equation (1), we get

$$\theta_2 = \frac{\mu_1}{\mu_2}(\alpha + \gamma)$$

Substituting this into second of eq.(1), we obtain

$$\mu_1\alpha + \mu_2\beta = (\mu_2 - \mu_1)\gamma \qquad ...(4)$$

Now, we use the approximations $\tan \alpha = \alpha$ and so on and also neglect the small distance δ.

Thus, $$\alpha = \frac{h}{u}, \beta = \frac{h}{\upsilon}, \text{ and } \gamma = \frac{h}{R} \qquad ...(5)$$

Finally, we substitute these into Eqn. (3) and divide out the common factor h. We obtain

$$\frac{\mu_1}{u} + \frac{\mu_2}{\upsilon} = \frac{\mu_2 - \mu_1}{R} \qquad ...(6)$$

We now apply the sign convention: u is –ve, υ is +ve and R is +ve. It gives us

$$\frac{\mu_2}{\upsilon} - \frac{\mu_1}{u} = \frac{\mu_2 - \mu_1}{R} \qquad ...(7)$$

This is the *Gaussian relation* for a single spherical surface. Though this relation is derived for a convex surface and for a real object and real image, it is equally valid for all other conditions, as shown in the following sections. If the first medium is air and the second medium is of refractive index μ, then $\mu_1 = 1$ and $\mu_2 = \mu$.

$$\frac{\mu}{\upsilon} - \frac{1}{u} = \frac{\mu - 1}{R} \qquad ...(8)$$

(b) Refraction at a Convex Surface Forms a Virtual Image

When the point object O lies on the principal axis close to the vertex of the convex refracting surface, it forms a virtual image.

In Δ^{le}OAC $\qquad \theta_1 = \alpha + \gamma$ and

In Δ^{le}IAC $\qquad \theta_2 = \beta + \gamma$

Using these in the Snell's equation, we get

$$\mu_1(\alpha + \gamma) = \mu_2(\beta + \gamma)$$

$$\mu_1\alpha - \mu_2\beta = (\mu_2 - \mu_1)\gamma$$

Substituting the values of α, β, and γ, and dividing throughout by h, we get

$$\frac{-\mu_2}{\upsilon} + \frac{\mu_1}{u} = \frac{(\mu_2 - \mu_1)}{R} \qquad ...(9)$$

According to sign convention both u and υ are negative and R is positive.

$$\therefore \qquad \frac{-\mu_2}{-\upsilon} + \frac{\mu_1}{-u} = \frac{(\mu_2 - \mu_1)}{R}$$

or

$$\frac{\mu_2}{\upsilon} - \frac{\mu_1}{u} = \frac{\mu_2 - \mu_1}{R} \qquad ...(10)$$

Refraction at a Concave Surface

Let O be the point object placed on the principal axis. The ray OA incident on the surface at angle θ_1 is refracted along AB at an angle θ_2.

The ray of light starting from O and travelling along the principal axis passes undeviated through vertex P.

The two refracted rays do not meet at any point but appear to come from point I on the principal axis. Therefore, I is the virtual image of the object O.

Referring to the in Δ^{le} OAC $\theta_1 = \gamma - \alpha$ and

In Δ^{le}IAC, $\quad \theta_2 = \gamma - \beta$

Using these in the Snell's equation (30), we get

$$\mu_1(\gamma - \alpha) = \mu_2(\gamma - \beta)$$

$$-\mu_1\alpha + \mu_2\beta = (\mu_2 - \mu_1)\gamma$$

Substituting the values of α, β, and γ, and dividing throughout by h, we get

$$\frac{\mu_2}{\upsilon} - \frac{\mu_1}{u} = \frac{(\mu_2 - \mu_1)}{R} \quad ...(11)$$

According to sign convention, the quantities u, υ and R are all negative.

$$\therefore \quad \frac{\mu_2}{-\upsilon} - \frac{\mu_1}{-u} = \frac{(\mu_2 - \mu_1)}{R}$$

$$\Rightarrow \quad \frac{\mu_2}{\upsilon} - \frac{\mu_1}{u} = \frac{\mu_2 - \mu_1}{R} \quad ...(12)$$

PRINCIPAL FOCI

We can use eqn. (12) to determine the first and second principal foci. F_1 is the position of the object for which the image lies at infinity, while F_2 is the position of the image for which the object lies at infinity.

First Principal Focus

If the object is placed at F_1 such that the refracted rays become parallel to the principal axis, $\upsilon = \infty$. The object distance u is denoted by f_1 and is called the *first principal focal length* of the refracting surface.

$$\frac{\mu_2}{\infty} - \frac{\mu_1}{f_1} = \frac{\mu_2 - \mu_1}{R}$$

$$\therefore \quad f_1 \frac{\mu_1 R}{\mu_2 - \mu_1} \quad ...(1)$$

The position of the axial point object whose image is formed at infinity is known as the *first principal focus* of the refracting surface.

Second Principal Focus

If the object is at infinity, then $u = \infty$ and $1/u = 0$. Let f_2 be the value of υ corresponding to $u = \infty$. Then

$$\frac{\mu_2}{f_2} = \frac{\mu_2 - \mu_1}{R} \quad ...(2)$$

$$\Rightarrow \quad f_2 = \frac{\mu_2 R}{\mu_2 - \mu_1}$$

This value of $\upsilon = f_2$ is termed as the *second principal focal* length and the position of the image corresponding to the object at infinity is called the *second principal focus* of the refracting surface.

It is easy to see that *the first focal length f_1 for the spherical refracting surface is not equal to the second focal length f_2*

In case of a convex surface R is positive and f_1 is negative and f_2 is positive. On the other hand, for a concave surface R is negative and hence f_1 is positive and f_2 is negative. When f_1 is positive the system is called *convergent*, and when f_1 is negative it is called *divergent.*

The following points are interesting and are to be remembered.

If *parallel light* is incident on a *convex surface*, the light *converges* at the *second focal point*, F_2. If the *convex surface renders* the incident *light into a parallel beam*, the light comes from the *first focal point*, F_2

The same applies to a concave surface also. In case of a convex surface, the second focal point F_2 is in the medium at right side while the first focal point F_1 is in air at left side to the surface. The reverse is true in case of a concave surface.

Various Relations Involving f_1 ad f_2

(i) The sum of the two focal lengths equals the radius of curvature.

$$f_1 + f_2 = \left[\frac{-\mu_1}{\mu_2 - \mu_1}\right] R + \left[\frac{\mu_2}{\mu_2 - \mu_1}\right] R = R \qquad ...(3)$$

(ii) We see that using relation (2) into spherical mirror equation, we get

$$\frac{\mu_1}{u} - \frac{\mu_2}{\upsilon} = \frac{\mu_1}{f_2} \qquad ...(4)$$

This is known as *Gauss' formula for a spherical surface.*

(iii) We may write

$$\frac{f_1}{\mu_1} = \frac{-R}{\mu_2 - \mu_1}, \frac{f_2}{\mu_2} = \frac{R}{\mu_2 - \mu_1}$$

Adding, $$\frac{f_1}{\mu_1} + \frac{f_2}{\mu_2} = 0$$

$$\Rightarrow \qquad \frac{f_1}{f_2} = -\frac{\mu_1}{\mu_2} \qquad ...(5)$$

As μ_1 and μ_2 are always positive and unequal, f_1 and f_2 are always of opposite sign and are unequal. The ratio of focal lengths f_1/f_2 is equal to the ratio μ_1/μ_2 of the corresponding refractive indices.

(iv) We may write $\dfrac{\mu_2}{\upsilon} - \dfrac{\mu_1}{u} = \dfrac{\mu_2 - \mu_1}{R}$

$$\Rightarrow \quad \frac{\mu_2 R}{\upsilon(\mu_2 - \mu_1)} - \frac{\mu_2 R}{u(\mu_2 - \mu_1)} = 1$$

$$\therefore \quad \frac{f_2}{\upsilon} + \frac{f_1}{u} = 1 \qquad \text{...(6)}$$

APLANATIC POINTS OF A SPHERICAL SURFACE

A spherical refracting surface forms a point image of an axial point object only in the case of paraxial rays. When this condition is not satisfied, the refracted rays will not have a common point of intersection. The image of the axial point object will not be a point image. This departure from the ideal point-image is termed as *spherical aberration.* However, this defect may be overcome by the use of an aplanatic surface. An image-forming surface, which is free from spherical aberration and coma, is known as an *aplanatic surface.*

It is a spherical surface and is, with respect to two conjugate points, such that all the rays originating from one of the two points are reflected or refracted through the other point. Therefore, the surface forms a geometrical point image of a point object on its axis, however big the aperture of the surface may be. These conjugate points are known as *aplanatic foci* or *aplanatic points.*

A spherical surface APB, having C as the centre of its curvature and P as its pole. Let the refractive index of the medium of the sphere be m_1 and that of the medium outside the sphere be m_2 where $m_1 > m_2$. Consider any axial wide-angled ray OA from an axial point object O striking the surface at point A. The refracted ray bends away from the normal and the virtual image is formed at point I. It will now be shown that, provided object O is so placed that OC = (m_2/m_1) R, then all rays I originating from 0 and striking the surface APB will seem to come from I. That is, I is a perfect point image of the point object 0 and free from spherical aberration.

Referring to the ray OA is incident on the sphere at A and is bent away from the normal and is refracted along AQ. Another ray OC, incident normally at P, passes straight without any deviation. The two rays AQ and OCP appear to diverge from I, which is the virtual image of O.

Now $\angle$ OAC = i, $\angle$ NAQ = $\angle$ IAC = r, $\angle$ AOC = α,

and $\angle$ AIC = β

In Δ^{le}AOC,

$$\frac{\sin i}{\sin \alpha} = \frac{CO}{CA} = \frac{CO}{R}$$

The object is so placed that CO = $\frac{\mu_2}{\mu_1}R$.

$$\frac{\sin i}{\sin \alpha} = \frac{\mu_2 R}{\mu_1 R} = \frac{\mu_2}{\mu_1} \qquad ...(1)$$

From Snell's law $\frac{\sin i}{\sin r} = \frac{\mu_2}{\mu_1}$ ....(2)

$$\frac{\sin i}{\sin r} = \frac{\sin i}{\sin r} \qquad ...(3)$$

$$\sin \alpha = \sin r$$

$\therefore$ $i = r$

In the Δ^{le}IAO, exterior $\angle$AOC = β + $\angle$IAO

$\therefore$ $\alpha = \beta + r - i.$

But $\alpha = r$

$\therefore$ $\beta = i$

From Δ^{le}IAC, $\frac{\sin \beta}{\sin r} = \frac{CA}{CI}$

Since β = i, and CA = R, $\frac{\sin i}{\sin r} = \frac{R}{CI}$ ...(4)

Using eqn. (46) into the above relation (48), we get

$$\frac{\mu_2}{\mu_1} = \frac{R}{CI}$$

$$\therefore \quad CI = \frac{\mu_1 R}{\mu_2} \qquad ...(5)$$

Thus, if CO = $\frac{\mu_2}{\mu_1}R$, then the corresponding image lies at a distance CI such that CI = $\frac{\mu_2 R}{\mu_2}$ and vice versa. Hence, the distance CI is independent of the position of A. In other words, all rays from point object O, however large an angle they make with the axis, appear to diverge from point I after emerging from the boundary between the two media. Therefore, there is no spherical aberration for this particular position of the object. Point O and I are the aplanatic points of the refracting surface.

Such aplanatic surfaces are made use of in microscope objectives where it is essential that as wide a pencil of light as possible from each point of the extremely small object should enter the objective; otherwise the greatly magnified image would be too faintly illuminated to be seen distinctly.

The objective is so designed that the centre of curvature O of the front surface C coincides with the aplanatic point of the outer surface D. Hence, if a point object is placed at O, then rays from it enter the lens normally and suffer no refraction or deviation at the first surface. Therefore, the refraction at the outer surface D is without any spherical aberration.

SOME LAWS OF RAY OPTICS

Optics is regarded by many as a branch of physics, which has outlived its purpose. This is not true. Since the invention of camera, and subsequently movies and TV, optics has been playing a key role in the entertainment industry. Reflection, refraction, and dispersion are fundamental processes of optics. These can be understood using either ray concept or wave concept.

The transmission and storage of images is central to many fields, such as remote sensing, and space exploration. Optics has regained its status in the eyes of public with the advent of optical fibres and their application in communications. The ray concept to explain the principles and working of optical systems and instruments.

Ray optics is based mainly on the following three simple laws stated below:

SPHERICAL ABERRATION

A lens may be regarded as made up of a large number of prisms, of increasing angles from the centre to outward in case of a convex lens and of decreasing angles in case of concave lens. A ray of light falling on a prism of larger angle is deviated more towards the base of the prism than that falling on a prism of small angle. Therefore, peripheral light rays passing through a lens farther away from the axis are refracted more and come to focus closer to the lens. Paraxial rays passing through the lens close 10 the axis are refracted less and come to focus farther from the lens. Therefore, rays passing through different zones of a lens surface come to different foci. An image formed by paraxial rays will be surrounded by a diffuse halo formed by peripheral rays and consequently the image is blurred. This phenomenon is known as *spherical aberration.* In other words, if we consider a lens with *spherical surfaces*, and an object point P on *the axis*, then the rays near the axis (paraxial rays) and the rays near the margin of the lens (*marginal rays*) form images at different points P_a and P_m. The distance P_aP_m is called *longitudinal spherical aberration.* In a transverse plane at the paraxial image point P_a the image is not a point, but is a circle of some radius. This radius gives the *transverse spherical aberration. Spherical aberration* is due to large aperture of the lens and concerns the image of an axial object point.

The presence of spherical aberration in the image formed by a single lens is illustrated in Fig. 9.2. O is a point object on the axis of the lens and I_p and I_m are the images formed by the paraxial and marginal rays respectively. It is clear from the Fig. 9.2 that the paraxial rays of light form the image at a longer distance from the lens than the marginal rays. The image is not sharp at any point on the axis. However, if the screen is placed perpendicular to the axis at AB, the image appears to be a circular patch of diameter AB. At positions on the two sides of AB, the image patch has a larger diameter. This patch of diameter AB is called the *circle of least confusion*, which corresponds, to the position of the best image. The distance I_n I_p measures the *longitudinal spherical aberration.* The radius of the circle of least confusion measures the *lateral spherical aberration.* When the aperture of the lens is relatively large compared to the focal length of the lens, the cones of the rays of light refracted through the different zones of the lens surface are not brought to focus at the same point I_m and the axial rays come to focus at a farther point I_p. Thus, for an object point O on the axis, the image extends over the length I_m I_p. This effect is called *spherical aberration* and arises due to the fact that different annular zones have different focal lengths.

The spherical aberration produced by a lens depends on the distance of the object point and varies approximately as the square of the distance of the object ray above the axis of the lens. The spherical aberration produced by a convex lens is positive and that produced by a concave lens is negative.

LAW OF RECTILINEAR PROPAGATION OF LIGHT

It is a common sight that an object kept in the path of light coming from a point source produces a sharp shadow. Similarly, on misty mornings sunlight passing through the dense foliage of trees appears as light shafts. There are many such instances where light appears to exhibit rectilinear propagation.

The law of rectilinear propagation states that in an optically homogeneous medium light propagates in a straight line. A medium is said to be optically homogeneous if its refractive index is everywhere the same. This law is approximate and when light passes through very small openings, deviations from a straight line are observed.

LAW OF INDEPENDENCE OF LIGHT RAYS

Light rays do not disturb one another when they intersect. If several rays are passing through a medium simultaneously in different directions, then the path of any ray is the same as it would be if all others were absent. The intersection of rays does not hinder the rays from propagating independently of each other. Rays of light always preserve their individuality. The rays pass through each other simultaneously without mutual interaction. For example when we view an object, light rays passing in other directions does not obstruct the light that comes to us from the object.

LAW OF REVERSIBILITY OF PATH

If the path of a light ray is reversed, it will exactly retrace its path, irrespective of the number of reflections and refractions.

LENS ABERRATIONS

The departures of real images from the ideal images, in respect of the actual size, shape, and position, are called *aberrations*. In other words, an aberration is any failure of a mirror or a lens to behave precisely according to the simple *formulae* we have derived. Aberrations are only due to inherent shortcomings of a lens and not caused by faulty construction of the lens, such as irregularities in its surfaces. *They are inevitable consequences of the laws of refraction at spherical surfaces.* In this chapter, we shall consider aberrations present in image formed with monochromatic light. These aberrations occur if:

(a) The optical system has large aperture, or

(b) The rays make large angles with the principal axis, or

(c) The object has large size perpendicular to the axis.

In the ideal theory of image formation we always use the approximation sin θ = 0. This is not true if the angles involved are large. We may get a first estimate of the errors if we take $\sin\theta = \theta - \frac{\theta^3}{3!}$. The theory developed is then called the *'third order'* theory. For even greater accuracy θ^5 term may also be considered (fifth order theory). We shall not give these theories, but only refer to the results and indicate the steps needed to avoid or minimise the aberrations.

One of the basic problems of lenses is the imperfect quality of the images. The simple equations derived connecting object and image distances, focal length etc. are based on the assumption that the angles made by the rays of light with the axis are small *and paraxial approximation* may be made. In actual practice the objects are bigger and a lens is required to produce a bright and magnified image. We are therefore, required to take into consideration the wide-angle rays from the centre of the object and also the upper and lower parts of the object and falling near the top and bottom of the lens. These rays are known as *peripheral or marginal* rays. In general, peripheral rays of light do not meet at a single point after refraction through the lens. Secondly, the refractive index and hence the focal length of a lens is different for different wavelengths of light. For a given lens, the refractive index for violet light is more than that for red light. Thus, if the light coming from an object point is not monochromatic, the lens forms a number of coloured images. These images, even though formed by paraxial rays, are at different positions and are of different sizes.

REFLECTION AT PLANE SURFACES (MIRRORS)

Any smooth surface acts as a mirror. A mirror may be *plane* or *curved.* Mirrors were usually made in the past, by coating glass with silver. Nowadays, they are made by depositing in vacuum a thin film of aluminium on a polished surface. The reflecting film is protected by deposition of a thin layer of silicon monoxide or magnesium fluoride over it. The plane mirrors used as looking glass are coated at the back so that the reflecting layer is completely protected. The mirrors used for technological purposes are coated on the front surface so that losses in energy due to transmission through the substrate material are reduced.

Image of a Point Object

Image formation by mirrors involves only the law of reflection. A point object O located at a distance u in front (to the left) of a plane mirror, u is called the *object distance*. The ray OA is incident normally on the mirror and is reflected along its original path. When we extend the two reflected rays backward, they intersect at point I, at a distance υ behind the mirror.

To the observer the reflected rays appear to come from I lying behind the mirror. Thus, I is the image of the object point O. The image is *virtual* because no rays actually come from it. We call υ the *image distance*. Let us draw the line OI perpendicular to the mirror. The two triangles OAB and IAB are congruent, so O and I are at equal distances from the mirror, and u and υ have equal magnitudes. It means that the image point I is located exactly opposite the object point O and is as far behind the mirror as the object point is from the front of the mirror.

The object distance u is positive because the object point is on the incoming side of the reflecting surface. The image distance v is negative because the image point I is not on the left side of the mirror. The object and the image distances are related by

$$u = -\upsilon \qquad ...(1)$$

It is easy to see that the incident ray is deviated through an angle $6 = 180° - 2i$.

Object and Image

An object is anything from which light rays radiate. This light could be emitted by the object itself or reflected from it. We refer normally to two kinds of objects-a *point object*, which has no physical extent, and an *extended or distributed object*, which has a length, width and height.

When light rays proceeding from an object are reflected at a mirror, they *converge to* or *appear to diverge from* a position *different* from that of the original object and give the impression of object being there. The *apparent object* as seen by the observer is called the *image* of the object.

The image formed by an optical component may be real or virtual.

The eye sees a virtual image as well as it sees a real image. A *real image* contains light energy; the image can be received and seen on a screen. And also, the real image can be photographed by simply placing a photographic film (or plate) at the position of the image. Real images are *essential* for photography.

On the other hand, a virtual *image* cannot be received on a screen. The light rays never actually pass through the virtual image; hence, the image cannot be received on a screen. For the same reason, it cannot be recorded on a photographic plate placed at its position. Virtual image can be photographed only with the help of a converging optical system, which uses the virtual image as a *virtual object* and forms its real image.

IMAGE OF AN EXTENDED OBJECT

We now consider an *extended object* with finite size. For sake of simplicity, we consider a linear object such as an arrow with its height oriented parallel to the mirror. Image formation by mirror.

An arrow AB of height h. The image formed by such an extended object is an extended image; to each point on the object, there corresponds a point on the image. Two of the rays from B are shown. The ray BD is incident normally on the mirror and is reflucted along the path DB. The ray BC, incident on the mirror at an angle i, is reflected at an equal angle along CQ. These two (in fact *all*) rays from B appear to diverge from the image point B' after reflection. The image of the arrow is the line A'B', with height h'. Other points of the object AB have image points between A' and B'. The triangles ABC and A'B'C are congruent, so the object AB and image A'B' have the same size and orientation, and h = h'. Thus, the image is erect.

Lateral magnification, m, is defined as the ratio of image height to object height, h'/h. Thus,

$$m = \frac{h'}{h} \qquad ...(1)$$

In case of a plane mirror, m − + 1.

Note : The lateral magnification for a plane mirror is always + 1.

IMAGE REVERSAL

When we consider a three-dimensional object, a three-dimensional image is formed by the mirror. Three fingers of a hand form the object.

The images of two fingers that lie in a plane parallel to the mirror are just like their objects and are not reversed at all. The image of the finger that points toward the mirror is reversed front to back (Observe the arrows associated with the fingers).

Looking at it is also observed that a mirror reverses right and left. Then, obviously, the question arises in our mind–why does the mirror reverses right and left; and not up and down. Following the paths of rays, it is more accurate to say that a mirror reverses front to back than that it reverses right to left.

Effect of Rotation of Mirror on the Reflected Ray

When *a mirror rotates through an angle a, the reflected beam will move through double the angle, that is 2 α.*.

A ray AB being incident at an angle i on a mirror LM and reflected along BC. Now, the mirror is rotated through a small angle a (= $\angle$ LBL') because of which the ray AB is reflected along BC'. $\angle$CBC' is the angle through which the reflected ray gets displaced.

$$\angle CBC' = \angle ABC' - \angle ABC = 2(i + a) - 2i = 2\alpha.$$

Multiple Plane Mirrors

If an object is held between two or more plane mirrors, multiple virtual images are formed because of multiple reflections of light from the mirrors. The number of images formed depends upon the angle between the mirrors. Let the angle between the mirrors be a, then the number of images, N is given by

$$N = \frac{360°}{\alpha} - 1$$

For a pair of parallel mirrors $\alpha = 0°$, N is infinity and therefore, infinite number of images will be formed.

Properties of Images in Plane Mirrors

We now summarize the properties of images in plane mirrors.

(a) Image formed by a plane mirror is virtual and erect.

(b) The image is as far behind as the object is in front of it.

(c) The right side of the object is transformed into left side of the image.

(d) Magnification produced by a plane mirror is unity. Hence, the image is as large as the object.

(e) When a plane mirror is rotated through a certain angle, the reflected ray turns through twice the angle.

(f) If an object is held between two or more plane mirrors, multiple virtual images are formed. The number of images is given by

$$N = \frac{360°}{\alpha} - 1 \qquad ...(2)$$

where α is the angle between the mirrors.

LIGHT RAYS

The conception of energy travelling out from the source along rays can be used to explain the rectilinear propagation of light. We may define a light ray as a narrow stream of radiation that travels in a line, never diverging or converging. A light ray is represented by a line drawn in the direction in which the light energy is travelling.

We can think of a broad beam of light as a bundle of parallel rays. Ray concept was used to describe light long before its wave nature was firmly established. The ray model is highly useful in studying reflection, refraction and formation of images by mirrors and lenses.

ACHROMATISM OF TELESCOPE OBJECTIVE

A telescope is used for visual observation of distant objects. The important visible portion of the spectrum lies between red and blue colours. Therefore, the telescope objective must be achromatized between these colours, both for longitudinal and lateral aberrations.

Two lenses of different materials form an achromatic combination for two colours when they satisfy the following condition.

$$\frac{\omega_1}{f_1} + \frac{\omega_2}{f_2} = 0$$

or

$$\frac{\omega_1}{\omega_2} = -\frac{f_1}{f_2}$$

where f_1 and f_2 are the focal lengths of the lenses, and ω_1 and ω_2 are the dispersive powers of their materials. The focal length F of the combination is given by,

$$\frac{1}{F} = \frac{1}{f_1} + \frac{1}{f_2}$$

An achromatic telescope objective of a given focal length can be designed with the help of the above two equations.

(1) As ω_1 and ω_2 are positive quantities, f_1 and f_2 should have opposite signs. Therefore, one lens must be convex and the other concave.

(2) As the telescope objective is a converging lens system, the focal length f_1 of the convex lens should be smaller than the focal length f_2 of the concave lens. If f_1 is to be smaller than f_2, then ω_1 should be less than ω_2. Thus, the convex lens should be made of glass of smaller dispersive power. The dispersive power of crown glass is smaller than that of flint glass. Therefore, the achromatic combination must be made of a convex lens of crown glass and a concave lens of flint glass. The radii of curvatures R_1 and R_2 of the convex lens and R_1' and R_2' of the concave lenses can be calculated using the following equations.

$$\frac{1}{f_1} = (\mu_1 - 1)\left(\frac{1}{R_1} - \frac{1}{R_2}\right)$$

and $$\frac{1}{f_2} = (\mu_2 - 1)\left(\frac{1}{R_1'} - \frac{1}{R_2'}\right)$$

The relations give different sets of values of R_1 and R_2 (or of R_1' and R_2' for the same values of μ (or of μ_2). In practice, the objective is made plano-convex lens because it has minimum spherical aberration when convex side faces the incident light. Hence, we make one face of the concave lens plane (*i.e.* $R_2' = \infty$) and then calculate the radius of curvature R_1' of the other face. The surfaces of the lenses in contact are made of the same radius of curvature ($R_2 = R_1'$). The lenses are cemented together by Canada balsam to avoid loss of light by reflection at the interface. Using the value of R_2 the radius of curvature R_1 of the first face of the convex lens can then be determined.

ACHROMATISM OF A CAMERA LENS

The lenses used in cameras are achromatic doublets which consist of two lenses, one convex and the other concave, of different materials placed in contact so as to satisfy the following condition for two specified colours;

$$\frac{\omega_1}{f_1} + \frac{\omega_2}{f_2} = 0.$$

Usually, photographic plates are most sensitive between violet and green colours. Therefore, the camera lens doublet is achromatized for these two colours. In the achromatic doublet the equality of focal length is achieved for only two colours. In high quality cameras it is required to achieve equality of focal lengths for three or more colours. Then we use three or more component lenses of different materials placed in contact. Such combinations are called "apochromats". The materials, focal lengths and radii of curvature of these component lenses are so chosen that apochromat is free from spherical aberration, coma etc. monochromatic aberrations.

OIL IMMERSION OBJECTIVE OF HIGH POWER MICROSCOPE

The high-power microscopes have an overall magnification of about 500. The objective in these microscopes is required to produce a magnification of the order of 50. Hence the objective must satisfy the following conditions.

(i) The image produced by the objective should be free from chromatic aberration, spherical aberration and coma. The rest of the aberrations are unimportant because the field of view of a microscope is very small.

(ii) Sufficient light should enter through the lens as the object to be viewed is very small and is placed very close to the lens.

Oil immersion lenses make use of aplanatic surfaces for obtaining a very wide-angled pencil of light from each point of the object. If an object is placed at one aplanatic point of a surface, an image free from spherical aberration and coma is formed at the second aplanatic point. In case of a single spherical surface, these points are at distances R/μ and μ/R from the centre of curvature.

An object cannot be placed inside a glass lens. However, we can create an equivalent situation. The object is immersed in a film of oil, having the same refractive index as the material of the lens and the front surface of the lens is brought in contact with oil as shown in Fig. 9.37. This forms the oil-immersion objective lens of a microscope. The oil-immersion objective is in practice made of a combination of lenses as shown in Fig. 9.37. The lower-most lens L_1 that is close to the object O is a piano-hemispherical convex lens. The flat surface of the lens is kept in contact with cedar-wood oil whose refractive index is the same as that of L_1. The object O is immersed in the oil and for all practical purposes may be treated to be within the lens. The rays of light starting from O do not suffer refraction till they reach the spherical surface of the lens L_1. The depth of the oil is so adjusted that the distance of O is μ/R from the centre of curvature C of the lens L_1. Thus, O is an aplanatic point for the lens. Hence, all rays starting from it after refraction through the lens appear to come from the other aplanatic point I_1, where $CI_1 = \mu R$. The image I_1 is free from spherical aberration.

The position of the meniscus lens L_2 is so adjusted that the image I_1 lies at the centre of curvature of surface A of L_2. Therefore, the rays enter L_2 without deviation. I_1 is also the aplanatic point for the second surface B of L_2, which consequently forms a virtual image of I_1 at I_1. To sum up, the points O and I_1 are aplanatic points for the spherical surface S of the lens L_1. The points I_1 and I_2 are the aplanatic points for the second surface B of lens L_2.

Thus, the original wide-angled pencil of light starting from object O is changed into a narrow pencil diverging from point image I_2. The lenses L_3 and L_4 form the final image almost free from spherical aberration, chromatic aberration and coma.

SOLVED EXAMPLES

Example 1:

Two concave refracting surfaces of equal radii of curvature and refractive index 1.5 face each other in air as shown in Fig. 3.40. A point object is placed midway in between the centre and one of the vertices. What is the separation between the images of O formed by each refracting surface?

Solution:

We have for the image formed by the surface on the right side,

$$\mu_1 = \mu_2 = 1.5,$$

$$u = -R/2,\ R_1 = -R$$

$$\frac{\mu_2}{v} - \frac{\mu_1}{u} = \frac{\mu_2 - \mu_1}{R_1}$$

$$\frac{1.5}{v} - \frac{2}{-R} = \frac{1.50 - 1.00}{-R}$$

$$\therefore \quad v = -\frac{3R}{5}.$$

The image is at a distance of (R – 3R/5) = 0.4 R from the centre P towards the right side. For the surface on the left side,

$$\mu_1 = \mu_2 = 1.5 \text{ m} = 3R/2, \ R_2 = R$$

$$\frac{\mu_2}{v} - \frac{\mu_1}{u} = \frac{\mu_2 - \mu_1}{R_2}$$

$$\frac{1.5}{v} - \frac{2}{-3R} = \frac{1.50 - 1.00}{R}$$

$$\frac{1.5}{v} = \frac{1}{2R} + \frac{2}{3R} = \frac{7}{6R}$$

$$\therefore \quad v = \frac{9R}{7}$$

Hence, the image is at a distance of (9R/7 – R) = 2R/7 from the center towards the right hand side. The distance between the two images,

$$= 0.4R - 2R/7 = 0.114 \text{ R}.$$

Example 2:

A convex refracting surface of radius of curvature 15 cm separates two media of refractive indices 4/3 and 1.50. An object is kept in the first medium at a distance of 240 cm from the surface. Calculate the position of the image.

Solution:

$$\mu_1 = 4/3, \ \mu_2 = 1.50,$$

$$u = -240 \text{ cm},$$

$$R = 15 \text{ cm}$$

$$\frac{\mu_2}{v} - \frac{\mu_1}{u} = \frac{\mu_2 - \mu_1}{R}$$

$$\frac{1.50}{v} - \frac{4/3}{-240\text{cm}} = \frac{1.50 - 4/3}{15\text{cm}}$$

$$v = 270 \text{ cm}.$$

A real image forms in the second medium at a distance of 270 cm from the refracting surface.

Example 3:

The eye can be regarded as a single spherical refracting surface of radius of curvature of cornea 7.8 mm, separating two media of refractive indices. 1.00 and 1.34. Calculate the distance from the refracting surface at which a parallel beam of light will come to focus.

Solution:

$$\mu_1 = 1.00,\ \mu_2 = 1.34,$$

$$m = -\infty,\ R = 0.78 \text{ cm}$$

$$\frac{\mu_2}{v} - \frac{\mu_1}{u} = \frac{\mu_2 - \mu_1}{R}$$

$$\frac{1.34}{v} - \frac{1.00}{-\infty} = \frac{1.34 - 1.00}{0.78\text{cm}}$$

$$v = 3.075 \text{ cm.}$$

A real image forms in the second medium at a distance of 3.075cm from the refracting surface.

Example 4:

A point light source lies on the principal axis of a concave spherical mirror with radius of curvature 160 cm. Its vertical image appears to be at a distance of 70 cm from it. Determine the location of the light source.

Solution:

$$\frac{1}{u} + \frac{1}{v} = \frac{2}{R}$$

Here $v = 70$ cm,

$R = -160$ cm

$$\frac{1}{u} = \frac{2}{R} - \frac{1}{v}$$

$$\therefore \quad \frac{1}{u} = \frac{2}{-160\text{cm}} - \frac{1}{70\text{cm}} = -\frac{15}{560\text{cm}}$$

$$\therefore \quad u = -\frac{560}{15}\text{cm} = -37 \text{ cm}$$

The light is at a distance of 37 cm in front (to the left of the vertex) of the mirror.

Example 5:

A concave spherical surface of radius of curvatures 100 cm separates two media of refractive indices 1.50 and 4/3. An object is kept in the first medium at a distance of 30 cm from the surface. Calculate the position of the image.

Solution:

$\mu_1 = 1.50,\ \mu_2 = 4/3,$

$u = -30\text{ cm},\ R = -100\text{ cm}$

$$\frac{\mu_2}{v} - \frac{\mu_1}{u} = \frac{\mu_2 - \mu_1}{R}$$

$$\frac{4/3}{v} - \frac{1.50}{-30\text{cm}}$$

$$= \frac{4/3 - 1.50}{-100\text{cm}}$$

$v = -27.58$ cm.

Example 6:

Show that for refraction at a concave spherical surface (separating glass-air medium), the distance of the object should be greater than three times the radius of curvature of the refracting surface for the image to be real.

Solution:

$\mu_1 = 1.50,\ \mu_2 = 1$, u and R are negative.

$$\frac{\mu_2}{v} - \frac{\mu_1}{u} = \frac{\mu_2 - \mu_1}{R}$$

$$\frac{1}{v} - \frac{1.50}{-u} = \frac{1 - 1.50}{-R},$$

$$\frac{1}{v} + \frac{3}{2u} = \frac{1}{2R}$$

$$\therefore \quad \frac{1}{v} = \frac{1}{2R} - \frac{3}{2u}$$

For v to be positive $\frac{1}{2R} > \frac{3}{2u}$ or $u > 3R$.

Example 7:

A point source of light is located 20 cm in front of a cover mirror with f = 15 cm. Determine the position and character of the image point.

Solution:

$$\frac{1}{u}+\frac{1}{v} = \frac{1}{f}$$

Here $u = -20$ cm,

$f = 15$ cm.

$$\frac{1}{v} = \frac{1}{f}-\frac{1}{u} = \frac{1}{15\text{cm}}-\frac{1}{-20\text{cm}}$$

$$= \frac{35}{300\text{cm}} = \frac{6}{60\text{cm}}$$

As v is positive, the image is located behind (to the right side of the vertex of) the mirror. Hence, the image is *virtual.*

Example 8:

A boy 1.5 m tall stands in front of a vertical lane mirror. How tall must be a vertical mirror and high must its lower edge be above the floor if he is to be able to see his entire length? Assume his eyes are 10 cm below the to of his head.

Solution:

The boy JK standing in front of the mirror AB. It also shows the paths of rays that leaves his head and enter his eyes after relation from the mirror at point A; and the rays the leave his feet and enter his eyes reflected at the point B. The mirror need to be of length L = AB. From the Fig. 3.35, it is easy to see that

$$CG = \frac{1}{2}JG$$

$$\therefore \quad L = CD = CG + GD$$

$$\therefore \quad = \frac{1}{2}JG + \frac{1}{2}GK$$

$$= \frac{1}{2}(JG + GK) = \frac{1}{2}JK.$$

$$L = 1.5\text{ m}/2 = 0.75\text{ m}$$

The distance of the bottom edge of the mirror above floor

$= BE = GK/2 = (JK - JG)/2 = (1.5 - 0.1)/2 = 0.7$ m.

Example 9:

A convex surface of radius of curvature 40 cm separates two media of refractive indices 4/3 and 1.50. An object is kept in the first medium at a distance of 20 cm from the surface. Calculate the position of the image.

Solution:

$\mu_1 = 4/3,\ \mu_2 = 1.50,$

$u = -20$ cm, $R = 40$ cm

$$\frac{\mu_2}{v} - \frac{\mu_1}{u} = \frac{\mu_2 - \mu_1}{R}$$

$$\frac{1.50}{v} - \frac{4/3}{-20\text{cm}} = \frac{1.50 - 4/3}{40\text{cm}}$$

$v = -24$ cm.

Example 10:

A glass dumbbell of length 50 cm and refractive index 1.50 has ends of 5 cm radius of curvature. Find the position of the image formed due to refraction at one end only, when a point object is situated in air at a distance of 20 cm from the end of the dumbbell along the axis.

Solution:

$$\frac{\mu_2}{v} - \frac{\mu_1}{u} = \frac{\mu_2 - \mu_1}{R}$$

Here $\mu_1 = 1.50,\ \mu_2 = 1,$

$u = -20$ cm, $R = 5$ cm

$$\frac{1.5}{v} - \frac{1.00}{-20\text{cm}} = \frac{1.50 - 1.00}{5\text{cm}}$$

Since υ is positive, the image is formed 30 cm to the right of the vertex.

Example 11:

In the above example, if the object is 5 cm from the dumbbell, what is the position of the image?

Solution:

$$\frac{1.5}{v}-\frac{1.00}{-5cm}=\frac{1.50-1.00}{5cm}$$

$v = -15$ cm.

Therefore, the image is formed to the left of the vertex.

Example 12:

A small filament is at the centre of a hollow glass sphere if inner and outer radii 8 cm and 9 cm respectively. The refractive index of glass is 1.50. Calculate the position of the image of the filament when viewed from outside the sphere.

Solution:

For refraction at the first surface, $\mu_1 = 1$, $\mu_2 = 1.50$,

$m = -8$ cm, $R = -8$ cm

$$\frac{\mu_2}{v}-\frac{\mu_1}{u}=\frac{\mu_2-\mu_1}{R}$$

$$\frac{1.5}{v'}-\frac{1.00}{-8cm}=\frac{1.50-1.00}{-8cm}$$

$v' = -8$ cm.

It means that due to the first surface the image is formed at the centre of the sphere.

For the second surface,

$\mu_1 = 1.50$, $\mu_2 = 1$,

$u = -9$ cm, $R = -9$cm

$$\frac{\mu_2}{v}-\frac{\mu_1}{u}=\frac{\mu_2-\mu_1}{R}$$

$$\frac{1}{v}-\frac{1.50}{-9cm}=\frac{1-1.50}{-8cm}$$

$v = -9$ cm.

Hence, the final image is formed at the centre of the sphere.

Example 13:

A glass rod has ends. The refractive index of glass is μ. The object point O is at a distance 2R from the surface of larger radius of curvature.

The distance between the apexes of the ends is 3R. Show that the image point I is formed at a distance of $\frac{(9-4m)R}{(10m-9)(m-2)}$ *from the right hand vertex.*

Solution:

For the first surface $\frac{\mu_2}{v'} - \frac{\mu_1}{u} = \frac{\mu_2 - \mu_1}{R}$

$\mu_1 = \mu_2 = m,$

$u = -2R, R_1 = R$

$$\frac{\mu}{v'} - \frac{1}{-2R} = \frac{\mu - 1}{R}$$

or $\frac{\mu}{v'} = \frac{2\mu - 3}{2R}$

$$\therefore \quad v' = \frac{2\mu R}{2\mu - 3}$$

For the second surface,

$$u = -\left[3R - \left(\frac{2\mu R}{2\mu - 3}\right)\right]$$

Here $\mu_1 = m, \mu_2 = 1, R_2 = R/2$

$$\frac{\mu_2}{v} - \frac{\mu_1}{u} = \frac{\mu_2 - \mu_1}{R}$$

$$\frac{1}{v} + \frac{\mu}{\left[3R\left(\frac{2\mu R}{2\mu - 3}\right)\right]} = \frac{1 - \mu}{R/2}$$

$$\therefore \quad \frac{1}{v} = \frac{2 - 2\mu}{R} - \frac{\mu(2\mu - 3)}{(6\mu R - 9R - 2\mu R)}$$

$$\therefore \quad \frac{1}{v} = \frac{-10\mu^2 + 29\mu - 18}{R(4\mu - 9)}$$

$$\therefore \quad v = \frac{(9 - 4\mu)R}{(10\mu - 9)(\mu - 2)}$$

In this problem, the image will be real if the refractive index of glass is between 2 and 2.25. It will be virtual if the refractive index is less than 2 or is more than 2.25.

Example 14:

For refraction at a spherical surface of n = 1.5, R = 20 cm, evaluate the aberration for h = 5 cm and h = 7 cm.

Solution:

From Fq. (4) the computation is direct. For h = 5 cm.

$$SA = \frac{5^2}{2\times1.5\times1.5\times20} = 0.28 \text{ cm}$$

For h = 7 cm, then

$$SA = 0.28 \times \frac{7^2}{5^2} = 0.55 \text{ cm.}$$

Example 15:

An oil immersion objective and a meniscus lens are used in succession, each having n = 1.60. Calculate the semi-angles of the incident and emergent cone of rays.

Solution:

For the incident cone, $\tan \theta r \div (r/n) = n = 1.60$ $\theta = 58°$

For the intermediate cone, $\tan \theta' = r \div (rn) = 1/n = 0.667$ $\theta' = 33°42'$

In the meniscus lens θ' now becomes the incident cone. If θ_1'' is the emergent cone,

$$\tan \theta'' = \frac{1}{n^2}\tan\theta' = 0.261 \quad \theta'' = 14°36°.$$

Example 16:

Calculate the focal lengths of the crown and flint glass lenses that form an achromatic doublet of focal length 50 cm when placed in contact. The refractive indices for the C and F lines are given below:

Crown	$\mu_C = 1.515$	$\mu_F = 1.523$
Flint	$\mu_C = 1.644$	$\mu_F = 1.664.$

Solution:

$$\omega_1 = \frac{1.523-1.515}{\frac{1.523+1.515}{2}-1} = \frac{0.008}{0.519} = \frac{8}{519}$$

$$\omega_2 = \frac{1.664 - 1.644}{\frac{1.664 + 1.644}{2} - 1} = \frac{0.020}{0.654} = \frac{10}{327}$$

For achromatism

$$\frac{\omega_1}{f_1} + \frac{\omega_2}{f_2} = 0 \quad \text{or} \quad \frac{\omega_1}{f_1} = \frac{-\omega_2}{f_2}$$

$$\therefore \qquad \frac{1}{f_2} = \frac{\omega_1}{\omega_2 f_1} = \frac{8 \times 327}{519 \times 10 f_1}$$

$$\frac{1}{F} = \frac{1}{f_1} + \frac{1}{f_2} = \frac{1}{50}$$

$$\therefore \qquad \frac{1}{f_1} - \frac{8 \times 327}{512 \times 10 f} = \frac{1}{50}$$

or $$\frac{1}{f_1}\left[1 - \frac{8 \times 327}{519 \times 10}\right] = \frac{1}{50}$$

f_1 = 24.79 cm (crown glass, convex lens)

From equation (1) $f_2 = -\dfrac{519 \times 10 \times 24.79}{8 \times 327}$

= **– 49.19 cm.**

Example 17:

Calculate the focal lengths of a convex lens of crown glass (dispersive power 0.012) and a concave lens of flint glass (dispersive power 0.020) so that when placed in contact they form an achromatic converging combination of focal length 30 cm.

Solution:

For achromatism,

$$\frac{\omega_1}{f_1} + \frac{\omega_2}{f_2} = 0$$

Here, ω_1 = 0.012 and ω_2 = 0.020

$$\frac{0.012}{f_1} + \frac{0.020}{f_2} = 0 \qquad \therefore \quad \frac{1}{f_2} = -\frac{0.6}{f_1}$$

Also $$\frac{1}{F} = \frac{1}{f_2} + \frac{1}{f_2} \quad \text{Bit } F = +30 \text{ cm}$$

$$\frac{1}{30\text{cm}} = \frac{1}{f_1} - \frac{0.6}{f_1} \quad f_1 = 30 \text{ cm} \times 0.4 = 12 \text{ cm.}$$

$$\frac{1}{f_2} = -\frac{0.6}{12\text{cm}} \quad f_2 = -20 \text{ cm.}$$

The focal length of the convex lens is 12 cm and the focal length of the concave lens is 20 cm.

Example 18:

Two glasses have dispersive powers in the ratio 2:3. These glasses are to be used in the manufacture of an achromatic objective of focal length 20 cm. What are the focal lengths of the lenses?

Solution:

For an achromatic objective,

$$\frac{\omega_1}{f_1} + \frac{\omega_2}{f_2} = 0 \qquad \therefore \frac{1}{f_2} = -\frac{\omega_1}{\omega_2 f_1}$$

Also $$\frac{1}{F} = \frac{1}{f_1} + \frac{1}{f_2} = \frac{1}{50}$$

Here $$\frac{\omega_1}{\omega_2} = \frac{\omega_1}{\omega_2}, \qquad \text{and } F = 20 \text{ cm}$$

$\therefore$ $$\frac{1}{f_2} = -\frac{2}{3f_1}$$

$$\frac{1}{20} = \frac{1}{f_1} + \frac{1}{f_2}$$

$$= \frac{1}{f_1} - \frac{2}{3f_1} = \frac{1}{3f_1}$$

$$f_1 = 203 = \mathbf{6.67 \text{ cm.}}$$

Example 19:

The dispersive powers of crown and flint glass are 0.015 and 0.030 respectively. The refractive indices for the mean rays are 1.52 and 1.65. If one of the surfaces of the flint glass lens is plane, calculate the radii of curvature of the other surfaces of the two lenses which form an achromatic doublet of focal length 80 cm.

Solution:

Given that $\omega_1 = 0.015$,

$\omega_2 = 0.030$,

and $\mu_1 = 1.52$,

$\mu_2 = 1.65$

$$\frac{\omega_1}{f_1} + \frac{\omega_2}{f_2} = 0 \qquad \therefore \frac{1}{f_2} = -\frac{\omega_1}{\omega_2 f_1}$$

$$\therefore \qquad \frac{1}{f_2} = -\frac{0.015}{0.030 f_1} \quad \therefore \frac{1}{f_2} = -\frac{1}{2f_1}$$

But $\frac{1}{2f_1} = \frac{1}{f_1} + \frac{1}{f_2} = \frac{1}{80}$

or $\frac{1}{f_1} - \frac{1}{2f_1} = \frac{1}{80}$ or $f_1 = 40$ cm

$\therefore \qquad f_2 = -2f_1 = -80$ cm.

For the flint glass lens,

$$\frac{1}{f_2} = (\mu_2 - 1)\left(\frac{1}{R_1'} - \frac{1}{R_2'}\right)$$

Here, $R_2' = \infty \;\therefore\; -\frac{1}{80\text{cm}} = (1.65 - 1)\left(\frac{1}{R_1'}\right)$

or $R_1' = -52$ cm.

For the crown glass lens, $R_2 = R_1' = -52$ cm and

$$\frac{1}{f_1} = (\mu - 1)\left(\frac{1}{R_1} - \frac{1}{R_2}\right) \;\therefore\; \frac{1}{40\text{cm}} = (1.52 - 1)\left(\frac{1}{R_1} - \frac{1}{-52\text{cm}}\right)$$

$\therefore \; R_1 = 34.67$ cm.

For the crown glass lens, $R_1 = 34.67$ cm, $R_2 = -52$ cm

For the flint glass lens, $R_1' = -52$ cm, $R_2' = \infty$.

Example 20:

It is desired to make a converging achromatic lens of mean focal length 30 cm by using two lenses of materials A and B. If the dispersive powers of A and B are in the ratio 1:2, find the focal length of each lens.

Solution:

Given that $\frac{\omega_1}{\omega_2} = \frac{1}{2}$

We know that $\frac{\omega_1}{f_1} + \frac{\omega_2}{f_2} = 0 \therefore \frac{1}{f_2} = -\frac{\omega_1}{\omega_2 f_1} \therefore \frac{1}{f_2} = -\frac{1}{2f_1}$

$$\frac{1}{F} = \frac{1}{f_1} + \frac{1}{f_2} = \frac{1}{30}$$

or $$\frac{1}{f_1} - \frac{1}{2f_1} = \frac{1}{30} \quad \therefore f_1 = 15 \text{ cm}$$

From the equation (1) $f_2 = -2f_1 = -2 \times 15$ cm $= -30$ cm.

EXERCISES

1. What do you mean by aplanatic points and aplanatic surface? What are the advantages of aplanatic surfaces?
2. Obtain an expression for refraction at a single spherical surface. Hence derive the lens-maker's formula.
3. Derive an expression connecting the object and image distances, when refraction of light takes place at a spherical surface separating two media.
4. Discuss refraction at a convex surface when the image is virtual. Show that for air medium
5. Prove that in the case of refraction at a concave surface

   $$\frac{\mu}{\upsilon} - \frac{1}{u} = \frac{u-1}{R}$$
6. Discuss refraction at a convex surface when the image formed is real. Derive the necessary formula.

   $$\frac{\mu_2}{\upsilon} - \frac{\mu_2}{u} = \frac{u_2 - u_1}{R}$$
7. Explain the terms first principal focus and second principal focus and show that

   $f_1\upsilon + f_2\upsilon = u\upsilon$
8. Calculate the transverse and the longitudinal magnifications for refraction at a spherical surface.
9. Derive Helmholtz's relation as used for a number of coaxial refracting surfaces.

10. A light spot is 150 cm away from a convex spherical mirror with radius of curvature 72 cm. Compute the image distance.

11. An object is placed before a concave mirror normal to its principal axis so that its magnification is 1.2. After the object has been moved further from the mirror by 25 cm, the magnification becomes 0.4. Calculate the focal length and radius of curvature of the mirror.

12. A convex refracting surface of radius of curvature 50 cm separates two media of refractive indices 1.33 and 1.50. An object is placed at a distance of 100 cm in the first medium. Calculate the position of the image.

13. A convex refracting surface of radius of curvature 20 cm separates two media of refractive indices 1.33 and 1.50. An object is kept at a distance of 240 cm in the first medium. Calculate the position of the image.

14. A glass dumbbell of length one metre and refractive index 1.5 has ends of radius of curvature 5 cm. Calculate the position of the image due to refraction at one end only, when the object is at a distance of 50 cm from one end.

15. Obtain the condition for the formation of a real image when paraxial rays of light are refracted at a concave spherical water-air interface. Radius of curvature of the surface is R.

16. A small filament is at the centre of a hollow glass sphere of inner and outer radii 4 cm and 4.5 cm respectively. The refractive index of glass is 1.50. Calculate the position of the image of the filament when viewed from outside the sphere.

3

APPLICATION OF SOME METHODS IN OPTICAL SYSTEMS

INTRODUCTION

Let us consider the example of a double convex lens (Fig. 1). When a ray passes through a lens, refraction occurs twice, at each curved surface, and translation occurs once, between the two curved surfaces. Hence, we require two refraction matrices $[R_1]$ and $[R_2]$ and one translation matrix, $[T_{21}]$. Multiplication of these matrices, written from right to left, leads to the *system matrix*, [S]. The system matrix indicates how a ray of light is affected in passing through the optical system. Thus,

$$[R_2]\ [T_{21}]\ [R_1] = [S] \qquad \text{...(1)}$$

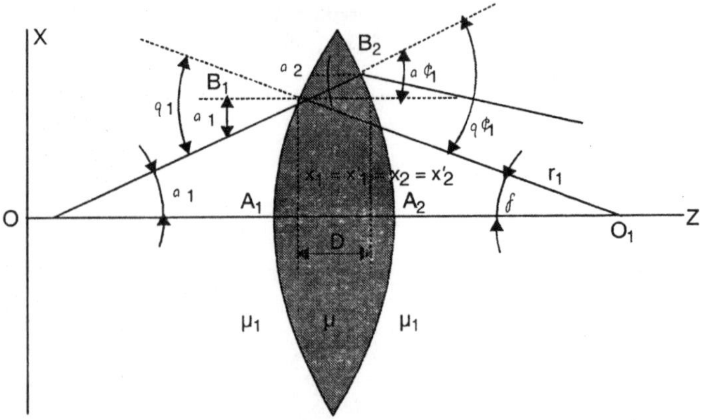

Fig. 1.

The passage of the ray through the first refracting surface is described by

$$\begin{pmatrix} \lambda_1' \\ x_1' \end{pmatrix} = \begin{pmatrix} 1 & -P_1 \\ 0 & 1 \end{pmatrix} \begin{pmatrix} \lambda_1 \\ x_1 \end{pmatrix} \qquad ...(1)$$

where P_1 is the refracting power of the first surface. Similarly, the propagation of the ray from the first to the second refracting surface inside the lens is described by

$$\begin{pmatrix} \lambda_2 \\ x_2 \end{pmatrix} = \begin{pmatrix} 1 & 0 \\ D/\mu_1 & 1 \end{pmatrix} \begin{pmatrix} \lambda_1' \\ x_1' \end{pmatrix} \qquad ...(2)$$

The refraction at the second surface may be written as

$$\begin{pmatrix} \lambda_2' \\ x_2' \end{pmatrix} = \begin{pmatrix} 1 & -P_2 \\ 0 & 1 \end{pmatrix} \begin{pmatrix} \lambda_2 \\ x_2 \end{pmatrix} \qquad ...(3)$$

where P_2 is the power of the second refracting surface. Combining the above equations (1), (2) and (3) we get the relation between the characteristics of the ray incident on the lens and the ray refracted by the lens. Thus, we get

$$\begin{pmatrix} \lambda_2' \\ x_2' \end{pmatrix} = [R_2][T_{21}][R_1] \begin{pmatrix} \lambda_1 \\ x_1 \end{pmatrix}$$

or
$$\begin{pmatrix} \lambda_2' \\ x_2' \end{pmatrix} = \begin{pmatrix} 1 & -P_2 \\ 0 & 1 \end{pmatrix} \begin{pmatrix} 1 & 0 \\ D/\mu_1 & 1 \end{pmatrix} \begin{pmatrix} 1 & -P_1 \\ 0 & 1 \end{pmatrix} \begin{pmatrix} \lambda_1 \\ x_1 \end{pmatrix} \qquad ...(4)$$

or
$$\begin{pmatrix} \lambda_2' \\ x_2' \end{pmatrix} = [S] \begin{pmatrix} \lambda_1 \\ x_1 \end{pmatrix}$$

$$\therefore\ [S] = [R_1]\ [T_{21}]\ [R_1] = \begin{pmatrix} 1 & -P_2 \\ 0 & 1 \end{pmatrix} \begin{pmatrix} 1 & 0 \\ D/\mu_1 & 1 \end{pmatrix} \begin{pmatrix} 1 & -P_1 \\ 0 & 1 \end{pmatrix} \qquad ...(5)$$

$$= \begin{pmatrix} b & -a \\ -d & c \end{pmatrix}$$

where a, b, c, and d are known as *Gauss constants*. Three of the four constants are independent and usually a, b, and c are chosen as the independent constants.

The matrix $$[S] = \begin{pmatrix} b & -a \\ -d & c \end{pmatrix} \qquad ...(6)$$

is called *system matrix*.

It follows that,

$$a = P_1 + P_2 - Pi_1 P_2 D/\mu_1 \quad ...(7)$$

$$b = 1 - P_2 D/\mu_1 \quad ...(8)$$

$$c = 1 - P_1 D/\mu_1 \quad ...(9)$$

$$d = -D/\mu_1 \quad ...(10)$$

Generally, an optical system consists of a number of refracting surfaces. So we get translation and refraction matrices for every surface. By multiplying all the individual matrices in proper order we get the final system matrix.

Since the system matrix is the product of translation and refraction matrices and as the determinants of these matrices are unity, the determinant of the system matrix is also unity.

$$\text{Thus det } S = \begin{vmatrix} b & -a \\ -d & c \end{vmatrix} = bc - ad = 1. \quad ...(11)$$

MATRIX METHODS BY OPTICAL SYSTEMS

Optical systems are made up of a large number of refracting surfaces. In order to obtain the position of the final image due to an optical system, one has to calculate step by step the position of the image due to each surface and consider this image as an object for the next surface. Such a step by step analysis becomes lengthy and tedious. In order to solve such problems easily and more efficiently K.

Hallback has introduced in 1964 matrix methods in the study of geometrical optics. The matrix method is less cumbersome and above that, it is more amenable to computer use. We study in this chapter the application of matrix method to the case of lenses. In this method we determine the translation matrix and refraction matrix of the light ray in an optical system. By multiplying these matrices we arrive at the system matrix. Using system matrix, we can find out the properties of the system.

CARDINAL POINTS OF AN OPTICAL SYSTEM

The cardinal points of an optical system are:

(i) two principal or unit points,

(ii) two focal points, and

(ii) two nodal points.

The position and size of the image of an object placed in front of a coaxial system of lenses (number of lenses having a common principal axis) can be determined with the help of the *cardinal points*. The cardinal points and planes of an optical system are shown in Fig. 2.

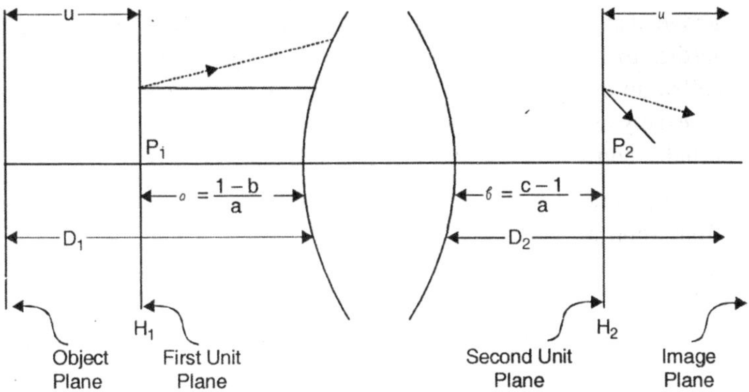

Fig. 2

P_1H_1 and P_2H_2 are the two unit planes. A ray emanating at any height from the first unit plane will cross the second unit plane at the same height.

PRINCIPAL OR UNIT POINTS

The principal or unit points (P_1 and P_2 in Fig. 7.4) are a pair of conjugate points on the principal axis for which linear transverse magnification is unity and positive. The planes passing through these points and perpendicular to the principal axis are called principal planes or unit planes.

Location of Unit Planes

The unit planes are two conjugate planes, one of which lies in object plane and the other in image plane. If a ray strikes the first unit planc (in object side) at a certain height, it emerges out from the second unit plane (on the image side) at the same height.

EYEPIECE

The magnifying power of simple microscope can be increased by decreasing the focal length of the lens. However, the focal length of a lens cannot be decreased beyond a certain limit. Moreover, the lens of small focal length has a smaller diameter because the curvature of the surface is large and the field of view is small.

Therefore, to increase the magnifying power, two separate lenses are used. The lens near the object is called the *objective*, which forms a real image of an object under examination. The lens used to enlarge this image further to form a, final image and which is then viewed by the eye is called an eyepiece or ocular. The objective of an ordinary microscope is usually an achromat with a magnification of about x_5 and an eyepiece usually consists of multiple lenses.

An optical instrument is required to produce a magnified image free from aberrations and a bright image covering a wide field of view. If a single lens is used as an eyepiece, the final image will suffer from spherical and chromatic aberrations. Another drawback is the small field of view, which becomes lesser and lesser as the magnification of the instrument is increased. The rays passing through the outer portions of the image are refracted through the peripheral portions of the eye lens and they cannot simultaneously enter the small aperture of the pupil of the eye placed close to the eye lens.

Hence, only that part of the image, which is nearer to the axis, will be seen. Therefore, the final image will cover a small field of view. The field of view will progressively decrease as the distance between the objective and ocular is increased. The distance is varied in order to increase the magnification. In other words, the greater the magnifying power of the instrument, the smaller the field of view. It becomes therefore expedient to place an additional lens in the eyepiece to cause all the rays from the image to enter the eye lens. This extra lens is called a *field lens*. The function of the field lens is to gather in more of the rays from the objective toward the axis of the eyepiece. The field lens and the eye lens together constitute an ocular or eyepiece. The two lenses are made and kept in such a way that their combination is achromatic and free from spherical aberrations. Two common eyepieces are the Huygens and the Ramsden types.

KELLNER'S EYEPIECE

It consists of two plano-convex lenses of equal focal lengths. The distance between the two lenses is equal to the focal length of either lens. The convex surfaces of both the lenses face the incident light.

(i) *Condition for Achromatism* : The distance between the two lenses for achromatism is

$$d = \frac{f_1 + f_2}{2} \quad ...(1)$$

In the present case $f_1 = f_2 = f$ (say)

$$\therefore \quad d = \frac{f + f}{2} = f \quad ...(2)$$

This is true for Kellner eyepiece. Thus, Kellner eyepiece satisfies the condition of achromatism.

(ii) *Condition of Minimum, Spherical Aberration* : The distance between the two lenses for minimum spherical aberration is

$$d = f_1 - f_2 \quad ...(3)$$

As $\quad f_1 = f_2 \quad d = 0.$

But the distance between the two lenses in Kellner eyepiece is f. Thus, the eyepiece does not satisfy the condition of minimum spherical aberration.

(iii) *Equivalent Focal Length* : The equivalent focal length of the eyepiece is given by

$$F = \frac{f_1 f_2}{f_1 + f_2 - d} = \frac{ff}{f + f - f} = f \quad ...(4)$$

The equivalent focal length of the eyepiece is thus equal to the focal length of either lens.

(iv) *Principal Points* : The distance of the first principal point P_1 from the lens L_1 is given by

$$\alpha_1 = \frac{dF}{f_2} = \frac{f\,f}{f} = f \quad ...(5)$$

That is, the first principal point lies at a distance/to the right of the field lens. The distance of the second principal point, P_2 from the eye lens L_2 is given by

$$\alpha_1 = \frac{dF}{f_2} = \frac{f\,f}{f} = f \quad ...(6)$$

That is, the second principal point lies at a distance of f to the left of the eye lens.

(v) *Focal Points* : The distance of the first focal point F_1 from the field lens L_1 is

$$\beta_1 = -F\left(1\frac{d}{f_2}\right) = -f\left(1-\frac{f}{f}\right) = 0 \quad ...(7)$$

That is, the first focal point coincides with the pole of the field lens.

The distance of the second focal point F_2 from the eye-lens L_2 is

$$\beta_2 = F\left(1\frac{d}{f_1}\right) = f\left(1-\frac{f}{f}\right) = 0 \quad ...(8)$$

That is, the second focal point coincides with the pole of the eye lens.

(vi) *Nodal Points :* As the lenses are in the air medium, the nodal points N_1 and N_2 coincide with the principal points P_1 and P_2 respectively.

POSITION OF CROSS-WIRES

An eyepiece when adjusted for normal vision forms the final image at infinity.

For this the objective should form the image at first focal plane. Hence, the cross-wire should be placed on the plane surface of the field lens, so that they may be in focus with the final image.

Merits and Demerits

1. The field of view of the eyepiece is very large. Therefore, it may be used in the microscope and high power telescopes.
2. As the eyepiece does not satisfy the condition of minimum spherical aberration, the defect is present.
3. The cross-wires are permanently marked on the field lens. Any scratch or dust particle present on the field lens also come into sharp focus and as a result the final image is spoiled.
4. The defect of distortion is produced.

HUYGENS EYEPIECE

In the Huygens eyepiece a converging beam enters the field lens and forms a virtual image before the eye lens. The need for a converging beam implies that this eyepiece does not act like a simple magnifier.

Construction

Huygens' eyepiece consists of two lenses having focal lengths in the ratio 3:1 and the distance between them is equal to the difference in their focal lengths. The focal lengths and the positions of the two lenses are such that each lens produces an equal deviation of the ray and the system is achromatic. This eyepiece is free from chromatic as well as spherical aberrations as it satisfies the two conditions simultaneously.

1. The lens combination acts as an achromatic system if $D = \dfrac{f_1 + f_2}{2}$

   where D is the distance between the lenses and f_1 and f_2 are the focal lengths of the two lenses.
2. The lenses produce equal deviation of the incident ray when the distance between the two lenses is equal to $(f_1 - f_2)$.

The field and the eye lenses used are plano-convex and are placed with their convex surfaces towards the incident ray. In this way, the total deviation due to the combination is divided into four parts which makes the combination to have minimum spherical aberration.

If spherical and chromatic aberrations are to be minimised simultaneously, the following condition is to be satisfied.

$$D = \frac{f_1 + f_2}{2} \qquad \text{Chromatic aberration}$$

$$D = f_1 - f_2 \qquad \text{Spherical aberration}$$

Combining the two conditions, we obtain

$$\frac{f_1 + f_2}{2} = f_2 - f_1$$

or

$$f_2 = 3f_1$$

$$D = 3f_1 - f_1 = 2f_1 \qquad \text{...(1)}$$

To satisfy the conditions for minimum chromatic and spherical aberrations, the focal length of the field lens should be three times the focal length of the eye lens and the distance between them should be equal to twice the focal length of the eye lens. Huygens' eyepiece is constructed on this principle.

Theory

The objective forms an image, which serves as a virtual object for the field lens. The field lens forms real inverte image $I'\,I'_1$. If this image is situated at the principal focus of the eye lens, then the final image is at infinity.

Equivalent Focal Length

The equivalent focal length of the eyepiece can be found as follows. If F is the equivalent focal length of the eyepiece, then it is given by

$$\frac{1}{F} = \frac{1}{f_1} + \frac{1}{f_2} - \frac{d}{f_1 f_2}$$

$$\frac{1}{F} = \frac{1}{f_1} + \frac{1}{f_2} - \frac{d}{f_1 f_2}$$

$$\therefore \qquad F\ 3f/2 \qquad \text{...(2)}$$

The equivalent lens lies behind field lens at a distance of

$$x = \frac{d \times F}{f} = \frac{2f \times 3f/2}{f} \qquad \text{...(3)}$$

In other words, the equivalent lens is at a distance of $3f - 2f = f$ behind the eye lens.

Position of Cross-Wires

It is noted above that the principal focus of the equivalent lens lies f/2 ahead of the eye lens. It is here that the image due to objective must be formed in order that the final image is at infinity.

The rays coming from the objective are, however, intercepted by the field lens, thus displacing the image to position I_1. It is here that the cross-wires should be placed.

The image I I_1 is at a distance of f/2 from the eye lens or 3f/2 from the field lens. Hence, for field lens $u = -3f/2$. Then

$$\frac{1}{u} + \frac{1}{\upsilon} = \frac{1}{f}$$

$$\therefore \qquad -\frac{2}{3f} + \frac{1}{\upsilon} = \frac{1}{3f}$$

$$\text{or} \qquad v = f \qquad \text{...(4)}$$

In other words, I I_1 lies midway between the two lenses and so fixes the position of the cross-wires or scale, if used. Since the image formed by the objective lies behind the field lens (instead of in the front) this eyepiece is sometimes referred to as a *negative eyepiece*. It may be noted that the image of the scale is formed by the eye lens whereas the final image is produced by both the lenses.

Hence the image and the scale would not be magnified equally and so the measurements will not be reliable. The image of the cross-wire or scale would have all the defects of an image formed by a single lens. Hence in instruments using Huygens' eyepieces, scale is not used except when magnification is low.

Merits and Demerits

(i) The Huygens' eyepiece is fully free from chromatic aberration because the distance between the lenses is equal to half the sum of their focal lengths.

(ii) Spherical aberration is also minimum because the distance between the two lenses is equal to the difference of their focal lengths.

(iii) The field of view of this eyepiece is smaller than that of Ramsden's eyepiece.

HUYGENS EYEPIECE POINTS

Position of Principal Points

The equivalent lens should be placed at a distance a from the field lens, which is given by

$$\alpha = \frac{d \times F}{f} = \frac{2f \times 3f/2}{f} = 3f \qquad ...(5)$$

Since the distance d between the field lens L_1 and the eye lens L_2 is 2f, the position of equivalent lens is $3f - 2f = f$, *i.e.*, it should be placed away from the eye lens, as shown by dotted line at P_1.

The first principal point P_1 lies at a distance $\alpha = 3f$ from the field lens. The second principal point P_2 lies at a distance β from the eye lens towards the field lens and is given by,

$$\beta = \frac{d \times F}{f} = \frac{-2f \times 3/2f}{3f} = -f \qquad ...(6)$$

Position of the Focal Points

The first focal point F_1 lies at a distance 3f/1 from the first principal point P_1, *i.e.*, at a distance $\left\{\frac{3f}{2} - f\right\} = \frac{f}{2}$ from the eye lens on the side of the field lens.

The second focal point lies at a distance 3f/2 from the second principal point P_2, *i.e.*, at a distance $\left\{\frac{3f}{2} - f\right\} = \frac{f}{2}$ from the eye lens away from the field lens.

RAMSDEN EYEPIECE

Ramsden's eyepiece consists of two plano-convex lenses each of focal length/separated by a distance equal to (2/3) f. The lenses are kept with their curved surfaces facing each other, thereby reducing spherical aberration. The field lens is a little larger than the intermediate image and is placed close to this image to allow as much light as possible to pass through it. The eye lens has a smaller diameter but carries out the actual magnification.

Theory

The objective forms the real inverted image I of a distant object. This serves as an object for the field lens, which gives rise to a virtual image I_1. I_1 in turn serves as an object for the eye lens, which gives the final image at infinity, because I_1 is made to lie at its principal focus.

Equivalent Focal Length

The equivalent focal length of the eyepiece can be found as follows. If F denotes the focal length of the equivalent lens, then

$$\frac{1}{F} = \frac{1}{f_1} + \frac{1}{f_2} - \frac{d}{f_1 f_2}$$

$$\therefore \qquad \frac{1}{F} = \frac{1}{f} + \frac{1}{f} - \frac{(2/3)}{f^2}$$

$$= \frac{2}{f} - \frac{2}{3f}$$

$$= \frac{4}{3f}$$

$$\therefore \qquad F = \frac{3f}{4} \qquad \text{...(1)}$$

The equivalent lens of focal length 3f/4 must be placed behind the field lens at a distance

$$\alpha = \frac{F \times d}{f} = \frac{\left(\frac{3}{4}\right)f \times \left(\frac{2}{3}\right)f}{f} = \frac{f}{2} \qquad \text{..(2)}$$

Thus, the equivalent lens lies between the field lens and the eye lens.

Position of the Cross-Wires

The cross-wires should be placed at the position of I. Now, the position of I relative to field lens can be found as follows. If the final image is to be formed at infinity, the image I should lie in the

focal plane of the equivalent lens. In other words, the distance AI should be equal to $F = \frac{3f}{4}$. Since $AL_1 = f/2$, $I_4 = f/4$. Therefore, the objective should produce the image at a distance of f/4 in front of the field lens. A fine scale may be placed here if it is desired to measure the size of the image. Since the scale and image would be magnified equally, the measurement would be trustworthy.

Merits and Demerits

(i) The field of view of this eyepiece is fairly wide.

(ii) It is not entirely free from chromatic aberration since the distance between the two lenses is not equal to half the sum of their focal iengths. However, chromatic aberration is minimised by using an achromatic combination both for the field lens and the eye lens.

(iii) Spherical aberration is minimised by using two plano-convex lenses thereby spreading deviation over four surfaces.

Ramsden's eyepiece is used practically in all instruments where measurements of the size of the final image are to be made. This eyepiece is sometimes referred to as *positive eyepiece* because a real image is formed by the objective in front of the field lens, which further acts as a real object for the eye lens.

POSITION OF THE PRINCIPAL PLANES

The first principal plane of the equivalent lens lies at P_1 behind the field lens at a distance a given by the relation

$$\alpha = \frac{F \times d}{f} = \frac{\left(\frac{3}{4}\right)f \times \left(\frac{2}{3}\right)f}{f} = \frac{f}{2} \qquad ...(1)$$

Thus, the equivalent lens is placed at P, in between the field lens F and the eye lens E at a distance f/2 behind the field lens. The point P_1 where the principal plane cuts the axis is the first principal point.

The second principal plane is at a distance β from the eye lens L_2.

$$\beta = \frac{d \times F}{f_1} = \frac{-\frac{2}{3}f \times \frac{3}{4}f}{f} = -\frac{f}{2} \qquad ...(2)$$

The second principal plane lies behind the eye lens L_1 at a distance f/2. The point P_2 where the principal plane cuts the axis is the second principal point.

Nodal Points

As the lens system is situated in air, the nodal points coincide with the principal points. Thus, P_1 and P_2 are the nodal points, N_1 and N_2 respectively.

Position of Focal Points

The equivalent focal length of the eyepiece is 3f/4. Hence, the first focal point F_1 lies at a distance 3f/4 from the first principal point P_1. That is at a distance of $\left\{\frac{3}{4}f - \frac{1}{2}f\right\} = \frac{1}{4}f$ to the left of the field lens L_1.

The second focal point F_2 lies at a distance 3f/4 from the second principal point P_2 *i.e.*, at a distance f/4 behind the eye lens L_2 away from the field lens.

GAUSS EYEPIECE

Gauss eyepiece is a modification of Ramsden eyepiece. The field lens and the eye lens are two plano-convex lenses of equal focal length and separated by a distance equal to two third of the focal length of either. To illuminate the field of view a glass plate G is held at an angle of 45° to the axis of the lens system. S is a source of light. Light reflected from G illuminates the field of view. The cross-wire C is kept at a distance 1/4f in front of the field lens. This eyepiece is used in the telescope of a spectrometer.

COMPOUND MICROSCOPE

A simple magnifying lens is not useful where large magnification is required. A highly magnified image must be produced in two stages. A compound microscope is used for the purpose. Around 1590, Janssen, a Dutch spectacle maker, invented the compound microscope. Galileo announced his invention of a compound microscope in 1610.

A compound microscope has two lenses, an objective and an eyepiece. The object AB is placed at a distance slightly greater than the focal length of the objective. The objective forms a real, inverted, magnified image A'B' at A'. The eyepiece is adjusted so that the distance of A' from it is less than its focal length. The eyepiece further magnifies the image A'B' and forms the final image PQ that is magnified virtual. Both the objective and the eyepiece of an actual microscope are highly corrected compound lenses with several optical elements, but for simplicity we show them here as simple thin lenses.

Magnifying Power : The overall angular magnification of the compound microscope is the product of two factors. The first factor is the lateral magnification m of the objective, which determines the linear size of the real image A'B' and the second factor is the angular magnification M of the eyepiece. When the microscope is in normal use, the final image PQ is formed at the near point, *i.e.* at D = 25cm from, the eye (Fig. 10.30).

Magnification of the objective $m = \frac{A'B'}{AB} = -\frac{\upsilon}{u}$

where, υ is the distance of the image A'B' from the objective and u is the distance of the object AB from the objective.

Magnification of the eyepiece $M_e = \left(1 + \frac{D}{f_e}\right)$ ...(1)

where, f_e is the focal length of the eyepiece.

Magnifying Power of the microscope

$$M = mM_e = -\frac{v}{u}\left(1 + \frac{D}{f_e}\right) = -\frac{v}{u}\left(1 + \frac{25cm}{f_e}\right) \quad ...(2)$$

Numerical Aperture

The brightness of the image is dependent on the amount of light gathered in by the objective. Ernst Abbe introduced the concept of *numerical aperture* to specify the light-gathering ability of an objective. It is given by,

$$N.A. = \mu \sin \theta_{max} \quad ...(3)$$

where μ is the refractive index of the medium adjacent to the objective lens, and θ_{max} is the semiangle of the maximum cone of light picked up by that lens. In other words, it is the angle made by a marginal ray with the axis.

The numerical aperture is independent of any refraction at plane surfaces at right angles to the axis of the instrument because, according to the law of refraction at plane surfaces

$$\mu \sin \theta = \sin \theta_1 = \text{constant} \qquad ...(4)$$

Thus, the value of N.A remains constant as the beam travels from one medium to another if they are separated by plane surfaces. The object is generally placed on a glass slide embedded in a material such as Canada balsam and protected by a thin cover glass. If the medium between the cover glass and the front lens of the objective is air, which is, of lower refractive index than the cover glass.

Then, $\quad$ N.A. $= \mu \sin \theta = \mu_1 \sin \theta_1$

For air $\mu_1 = 1$, $\theta_1 = 90°$ and μ for glass $= 1.5$

$\therefore \quad$ N.A. $= 1.5 \sin \theta = 1 \times \sin 90° = 1$

$\theta = 42°$ *i.e.,* most of the rays are lost by total internal reflection.

If we allow for some working distance between the cover glass and the objective, the maximum N.A is not reached and the maximum possible value will be 0.95.

If the space between the cover glass and the objective is filled with oil of the same refractive index as that of cover glass, the aperture is increased.

$$\text{N.A.} = \mu \sin 90° = 1.5 \times 1 = 1.5.$$

The angle of 90° is not reached due to the thickness of cover glass and the working distance. It has been found that the practical value for $\theta = 67.5°$

$$\text{N.A.} = \mu \sin 67.5° = 1.5 \times \sin 67.5° = 1.40$$

Hence, for a dry objective, $(\text{N.A})_{max} = 0.95$ and for an oil immersion objective, $(\text{N.A})_{max} = 1.40$.

The magnification and the numerical aperture are engraved on the barrel of the objective.

REFRACTING ASTRONOMICAL TELESCOPE

The simple astronomical telescope consists of two lenses (Fig. 10.32). The objective is a convex lens of long focal length and forms a real image of the distant object. The eyepiece is a convex lens of small focal length.

It may be noted that the optical system of a telescope is similar to that of a microscope. The objective lens forms a real and reduced image AB of the object. The eyepiece enlarges this image and forms a magnified virtual image at infinity. As the object viewed by the telescope lies far away, the first image is formed nearly at the second focal point of the objective lens. The telescope is in *normal adjustment* when the image is formed at infinity. If the final image formed by the eyepiece lies at infinity, the first image must also be at the first focal point of the eyepiece. The distance between the objective and eyepiece is therefore the sum of their focal lengths, F + f. To calculate the angular magnification M of the telescope, it is assumed that the eye is very close to the eyepiece. Moreover, the distance between the objective and the eyepiece is very small compared to the distance of the object from the objective or eyepiece. The angle a subtended at the unaided eye by the object can be taken as the angle subtended by the object at the objective.

$$M = \frac{\beta}{\alpha}$$

In the ΔABE, $\tan\beta = \frac{AB}{AE}$

or $\beta = \frac{AB}{AE}$

In the ΔOAB $\tan\alpha = \frac{AB}{OA}$

or $\alpha = \frac{AB}{OA}$

$$M = \frac{\beta}{\alpha} = \frac{\frac{AB}{AE}}{\frac{AB}{OA}} = -\frac{OA}{AE} = -\frac{F}{f} \qquad ...(1)$$

Thus, the magnification is equal to the ratio of the focal length of the objective and the eyepiece. The negative sign shows that the final image is inverted. To increase the magnification of the telescope, the objective should be of large focal length and the eyepiece should be of short focal length. By contrast, we may note that a microscope should have an objective of short focal length. An inverted image is not of any particular disadvantage for astronomical observations.

Jan Lippershey invented the first telescope in 1608. In the year 1609 Galileo designed a telescope with which he made astounding astronomical observations. Newton designed a reflecting telescope because of the difficulty in manufacturing lenses free from chromatic aberrations. An astronomical telescope is used to view heavenly bodies. The telescope that uses a lens as an objective is called a *Refracting Telescope*. However, many telescopes use a curved mirror as an objective. Such telescopes are known as *reflecting telescopes*.

Telescopes are used to view very distant objects, which subtend very small angle θ_1 at the eye. The telescope makes image of such objects so as to subtend an angle θ_1 at the eye, where $\frac{\theta_2}{\theta_1}$ is called the 'magnification', m. If f_0' is the focal length of the objective and f_E' that of the eyepiece, the angular magnification is given by $m = \frac{f_0'}{f_E'}$, whatever the eyepiece used.

Since f_E' cannot be too small in view of aberrations, large magnifying power requires large f_0', *i.e.*, objectives of very large focal lengths. The most serious error involved in telescope objectives is the *chromatic aberration.* Hence refracting telescopes must have achromatic doublets as objectives. All large telescopes are, in fact, of the reflecting type, where chromatic error is absent.

We further find that objectives of large focal length must also have large aperture.

This can be seen from the Helmholtz-Lagrange law. The *entrance pupil O* (aperture of the objective) and the *exit pupil* E (ultimately the eye) form a conjugate pair of planes. Hence, if ray from O at angler θ_1 emerges at E with angle θ_2, the transverse dimensions at O and at E must satisfy the relation

$$y_0 \tan\theta_1 = y_E \tan\theta_2.$$

Now $\frac{y_0}{y_E}$ is the appropriate ratio of diameters of the objective and the exit pupil $\left(\frac{d_0}{d_E}\right)$, and $\tan\theta_2/\tan\theta_2 \approx \frac{\theta_2}{\theta_1}$ is the magnification B| which in turn equals $\frac{f_0'}{f_E'}$.

$$\frac{d_0}{d_E} = m = \frac{f_0'}{f_E'}$$

Now, if the aperture d_0 of the objective is less than md_E, then full capacity to see details is not utilised (empty magnification). Also, if d_0 is greater than md_E then full capacity of the objective to show details is not utilised (empty resolution). Hence for 'normal' conditions,

$$do = m.d_E.$$

NEWTON'S TELESCOPE

The objective is a large concave spherical mirror made of speculum metal (alloy of copper and tin). A Ramsden eyepiece is used with this telescope. The rays from the distant star are reflected from the concave spherical mirror A of large aperture and long focal length. Then they are reflected to one side at right angles to the axis, by means of a small plane mirror B. The final image is seen through the eyepiece E.

REFLECTING TELESCOPE

In the reflecting telescope the objective lens is replaced by a concave mirror. In large telescopes this has many advantages. The objective of the telescope should have a large aperture if fine details and a bright image are required. In the case of a refracting telescope the objective is a combination of lenses. But objective of very large aperture cannot be manufactured.

Therefore, in very large telescopes the objective is a paraboloidal mirror instead of a converging lens. Mirrors are inherently free from chromatic and spherical aberrations. Spherical aberration in the case of a mirror can be avoided by the use of a paraboloidal mirror. The material of the mirror need not be transparent. Moreover, the mounting of a mirror is easier than that of a lens because the lens can be supported only at the rim. Because the image is formed in a region traversed by incoming rays, this image can be observed directly with an eyepiece by blocking part of the incoming beam. When a telescope is used for photography the eyepiece is removed and photographic film or an electronic detector is placed at the position of the real image. Most of the telescopes for research do not use an eyepiece.

The early reflecting telescopes used had objectives of speculum, an alloy of copper and tin. Later objectives made of glass silvered on its surface were used. At present objectives are made of aluminium coated Pyrex.

OTHER REFLECTING TELESCOPES

Herschel's telescope is similar to the Newton's telescope but the axis of the mirror A is inclined to the incident rays from the distant star. This avoids the use of the mirror. The final image is formed on one side and is seen through the eyepiece E. In *Gregory's telescope* the objective A has small central aperture. The rays from the distant star are reflected by the objective A. These rays form an image I at the point which lies between the focus and the centre of curvature of the concave mirror A. These rays after reflection from the mirror B form the final image I' which is viewed by the eyepiece E. I' is the secondary image of I.

Cassegrain's telescope has a large spherical nun-or A, having an aperture in the centre just similar to Gregory's telescope. The rays from the distant star after reflection from the mirror A fall on the convex mirror B and are allowed to converge at I. The final image is formed at I', which can be viewed through the eyepiece. The 500 cm Mount Polomar telescope is Cassegrain type.

Coude's telescope has a large paraboloidal mirror A. B is a convex mirror and C is a plane mirror. The rays from the distant star after reflection from the mirror A fall on the convex mirror B and after falling on the mirror C converge to a point I. The final image is formed at I which can be viewed through the eyepiece. The telescope is a combination of Newtonian and Cassegrainian telescopes. Moreover, the necessity of the aperture in the centre of the mirror A is avoided.

With reflecting telescopes, visual observation may be made in the Newtonian manner at the focus of the objective. Usually to direct photography for spectroscopic observations Cassegrain system is used. The photographs are taken by placing the plate at the principal focus of the objective. As very long exposures are required the telescope is mounted so as to turn about the polar axis by clockwork mechanism so that the star observed is always in the centre of the field.

There are many points in favour of reflecting telescopes:

1. There is no absorption of light as in thick lenses.
2. The mirrors are free from chromatic aberration.
3. In the case of paroboloidal mirror, there is no spherical aberration for beams parallel to the axis.
4. Mirrors can be constructed with considerably large diameters than lenses.

5. Mirrors can be easily mounted whereas lenses can be mounted only on the edges or rim.

CONSTANT DEVIATION SPECTROMETER

A constant deviation spectrometer has a collimator and a telescope fixed to rigid stand such that they are mutually perpendicular to one another. The prism P is placed on a table which can be rotate about a vertical axis with the help of a drum D. A is a spring which allows the lever B to move when the drum D is rotated. The drum D has a graduated scale, which is calibrated and the index C shows the wavelengths of the different lines directly.

To start with, the slit is illuminated with sodium light. The drum D is rotated such that the index C is on a mark showing the reading 5890Å. The prism P is placed on the table and rotated slowly with hand such that the yellow line D_1 of sodium light is seen in the field of view of the telescope and the line coincides with the centre of the cross wires. The prism is clamped in this position and sodium light is removed. The source whose wavelengths for different lines are to be determined, is placed before the slit. The drum D is rotated to bring the lines at the centre of the cross wires in turn and the corresponding readings as shown by the index C are read. Therefore, with the help of this spectrometer, wavelengths of spectral lines can be determined directly and in a short time. The scale is usually calibrated and standardized by using lines of known wavelengths from mercury light or a copper arc.

The constant deviation spectrometer is helpful in finding any impurity in different salts and also for radical analysis. The substance is allowed to emit radiation in a carbon arc and the wavelengths of different spectral lines are found. From the standard charts of wavelengths of the spectral lines of different substances, the nature of the radicals present in a salt can be found. If by any chance, the prism is disturbed, it can be reset by using a spectral line of known wavelength.

PULFRICH REFRACTOMETER

It is used to measure the refractive indices of solids and liquids. It consists of a right angled prism A having its two faces perfectly plane. One of the faces is horizontal and the other is vertical. The solid B whose refractive index is to be determined is taken having two faces cut perpendicular to one another. The faces are made perfectly plane.

The solid B is placed on the prism A. To bring perfect contact between the horizontal faces of A and B, a few drops of a liquid whose refractive index is greater than the solid is used. In the case of a liquid whose refractive index is to be determined, a glass cell is used.

Light is incident in a direction parallel to the horizontal surface so that light entering the prism A is incident at the critical angle c with the normal. Finally it emerges from the prism at an angle i.

Let the refractive index of the solid be μ_0. Here $\mu_0 > \mu$ and c is the critical angle

$$\sin c = \frac{\mu}{\mu_0} \quad ...(1)$$

also

$$\frac{\sin i}{\sin r} = \mu_0$$

$$r + c = \frac{\pi}{2} \quad r = \frac{\pi}{2} - c$$

$$\frac{\sin i}{\sin\left(\frac{\pi}{2} - c\right)} = \frac{\sin i}{\sin c} = \mu_0$$

$$\cos c = \frac{\sin i}{\mu_0} \quad ...(2)$$

Squaring and adding equations (1) and (2), we get

$$\sin^2 c + \cos^2 c = \frac{\mu^2}{\mu_0^2} + \frac{\sin^2 i}{\mu_0^2}$$

$$\mu = \sqrt{\mu_0^2 - \sin^2 i} \quad ...(3)$$

Knowing μ_0 and i, the value of μ can be calculated.

To measure i, the apparatus has a collimating telescope with a circular scale attached to it. The telescope is adjusted so that the cross wires lie on the dark edge of the field of view.

This gives the position of the minimum angle of emergence i, the material of the prism A has a refractive index of 1.74.

In modem instruments, the circular scale is calibrated in terms of the refractive index and readings can be read directly from the scale. In some cases, a table is provided with the instruments which give the value of μ corresponding to i.

ABBE REFRACTOMETER

Abbe refractometer is used to measure the refractive index of liquids. It consists of two prisms A and B. The prisms have angles 30°, 60° and 90°. The liquid whose refractive index is to be determined is placed between the two prisms. Light is reflected from the glass plate or a ground glass surface P into the prism A. The emergent light is received by the telescope. The rays or night at incident on the hypotenuse face of the prism at various angles if incidence. At a particular angle of incidence, the ray passes along the plane of the liquid and just passes into the second prism.

These are the rays, which have an angle of incidence equal to the critical angle. The theory is just similar to that of Pulfrich refractometer. The telescope is set to a position where the cross wires lie on the dark edge of the field of view. The reading is noted. The scale attached to the telescope is calibrated to give the refractive index of the liquid directly.

PRISM BINOCULARS

Prism binoculars consists of two astronomical telescopes each containing two totally reflecting right angled prisms. On this way, the distance between the objectives of the telescope is made larger than the distance between the two eyes, thus increasing the field of view and the stereoscopic effect pertaining to distance. Moreover the effective length is made 1/3 of the length of the telescope to be used for the same magnification and the final image is also erect.

The prism A is placed with its refractive edge horizontal while prism B is placed with its refracting edge vertical. In an astronomical telescope the final formed is inverted. The objective and the eyepiece are convex lenses. The prism A turns the image in the vertical direction upside down. The prism B turns the image in the horizontal direction. The image produced after reflecting is erect and magnified. The optical path is about three times the length of the binoculars. Therefore, it works as an astronomical telescope equal to three times its length.

OPTICAL INSTRUMENTS

The optical instruments can be broadly divided into two groups. One group forms *real images* of an object, which is projected onto a screen or photographic plate. The image can be viewed simultaneously by many observers. Projectors belong to this category. The other group forms *virtual images* of an object and only one observer can see the image. The virtual image formed by the instrument is transformed by the eye into a real image. Therefore, the optical instruments forming a virtual image are often called *optical aids*. The spectacles, microscopes and telescopes belong to this group. The reader is familiar with the *image forming action* of instruments like the eye, camera, microscope, telescope, epidiascope, etc. What follows will deal with two related questions:

(i) What factors decide the apertures of the various components, the position of the eye, the total Held of view, etc.?

(ii) Apart from the aberrations discussed in the last two chapters, what other factors are to be considered in designing a viewing instrument so that it may show best detail?

In effect most of our effort will be spent on the first question. The second question will be barely touched, because it would lead to more serious questions linked with the very nature of light, to be covered.

HUMAN EYE

Our eye is the most important optical instmment. Without eye, all other optical instruments

would have no value at all! A simplified diagram of the human eye. The main parts of the human eye are : cornea (C), iris (I), pupil (P), ciliary muscles (M), convex lens (called eye lens) (L), retina (R) and optic nerve (N). We will now describe the construction and working of the eye.

The front part of the eye is called cornea. It is made of a transparent substance and it is bulging outwards. The light coming from objects enters the eye through cornea.

Just behind the cornea is the iris (or coloured diaphragm). There is a hole in the middle of the iris which is called pupil of the eye. Thus, *pupil is a hole in the middle of the iris*. The pupil appears black because no light is reflected from it.

The eye-lens is a convex lens made of a transparent and flexible material like a jelly made of proteins. The eye-lens is held in position by ciliary muscles. The focal length of eye-lens (and hence its converging power) can changed by the action of ciliary muscles. In this respect an eye differs from a camera. The focal length of the convex lens used in a camera is fixed and cannot be changed but the focal length of the convex lens present inside the eye can be changed by the action of ciliary muscles,

The screen on which the image is formed in the eye is called retina. The retina is behind the eye lens and at the back part of the eye. The retina of an eye is just like the film in a camera. It should be noted that there is an eye-lid in front of the eye which is just like the shutter in a camera. When eye-lid is open, light can enter the eye but when eyelid is closed, no light enters the eye. The space between cornea and eye-lens is filled with a viscous liquid called 'aqueous humour'. And the space between eye-lens and retina is filled with another liquid called 'vitreous humour'. It is nearly spherical in shape and about 2.5 cm in diameter. It has a tough outer skin called *sclera*, which protects the eye and gives the necessary stiffness.

The sclera is lined inside with vascular tissue, which consists of blood vessels feeding the eye. At the front of the eye, the sclera extends into a thin convex transparent membrane called the *cornea*, which acts as an entrance lens for the eye. Back of the cornea is the iris, which is a pigmented muscular ring. The iris can be of different colours, which explains why people have blue, green, brown or black eyes. The iris contains an aperture with variable diameter called the *pupil*, which opens and closes to adapt to changing light intensity. Behind the iris is a *crystalline lens*, which is a biconvex lens made by Nature herself. The lens contains a fibrous jelly, hard at the centre and progressively softer at the outer portions. The crystalline lens is held in place by ligaments that attach it to the ciliary muscle, which encircles it.

The chamber in front of the lens and behind the cornea contains a liquid called *aqueous humor*. Behind the lens, the eye is filled with a thin water jelly called *vitreous humor*. The indices of refraction of both the aqueous humor and vitreous humor are about 1.336. The crystalline lens, while not homogeneous, has an average index of 1.437. Most of the refraction of light entering the eye occurs at the outer surface of the cornea. The chamber holding the vitreous humor is lined with a sensory layer called the *retina*.

The retina has the shape-of a hemisphere and contains light receptors called *rods and cones*. The human eye has a total of 125 million rods and 6.5 million cones. They sense the image and transmit it via the *optic nerve* to the brain. At the point where the optic nerve enters the eye, there are no rods or cones and this portion of the retina is called the *blind spot*.

In contrast, vision is most acute in a small central region at the axis of the eyeball called the *fovea centralis*. Here, cones responsible for colour vision are concentrated. The remaining areas of the retina are occupied mainly by rods.

Refraction at the cornea and the surfaces of the lens produces a real and inverted image of the object on the retina. The optic nerve sends a signal to the brain, which makes the corrections necessary for us to see objects in their natural positions.

For an object to be seen sharply, the image must be formed exactly at the location of the retina. The eye adjusts to different object distances by changing the focal length f of its lens; the lens-to-retina distance does not change. The focal length of the eye is adjusted by varying the radii of curvature of the crystalline lens, with the help of ciliary muscle. This process is called *accommodation*.

The eye accommodates involuntarily, but its ability to accommodate is not unlimited. We describe the limited ability of the eye in terms of the *far point* and the *near point* of the eye. The far point of a normal eye is infinity. The ciliary muscle is fully relaxed in this position. The nearest point on which the eye is focussed with the ciliary muscle fully contracted is the near point of the eye. The minimum distance at which an eye can see objects distinctly and without getting tired is called the *normal viewing distance or normal distance of distinct vision* (NDDV); it is 25 cm for a normal eye.

The range of accommodation gradually diminishes with age because the crystalline lens grows through a person's life and the ciliary muscles are less able to distort a larger lens. For this reason, the near point gradually recedes, as one grows older. This recession of the near point is called *preshyopia*.

WORKING OF THE EYE

The light rays coming from the object kept in front of us enter the pupil of the eye and fall on the eye-lens. The eye-lens is a convex lens, so it converges the light rays and produces a real and inverted image of the object on the retina. The image formed on the retina is conveyed to the brain by the optic nerve and gives rise to the sensation of vision. Actually, the retina has a large number of light-sensitive cells. When the image falls on the retina then these light-sensitive cells get activated and generate eiectrical signals. The retina sends these electrical signals to the brain through the optic nerve and gives rise to the sensation of vision. Although, the image formed on the retina is inverted, our mind interprets the image as that of an erect object. A small region of the retina where the optic nerve enters the eye-ball, is insensitive to light and it is called blind spot. Thus, *as far as physics is concerned, the eye consists of a convex lens (called eye lens) and a screen (called retina). The eye-lens forms a real image of the objects on the retina of the eye and we are able to see the objects.*

THE FUNCTION OF IRIS AND PUPIL

The iris automatically adjusts the size of the pupil according to the intensity of light received by the eye. If the amount of light received by the eye is great (as during day light), then pupil contracts (becomes small) and reduces the amount of light entering the eye. On the other hand, if the amount of light received by the eye is small (as in a dark room or during a, dark night), the pupil expands (becomes bigger) and more light enters the eye.

Thus, we can say that *the iris regulates the amount of light entering the eye by adjusting the size of the pupil.* Another way of saying this is that *the amount of light entering the eye can be controlled by the pupil.* If the intinsity of the outside light is low, then the pupil expands to allow more light to enter the eye. On the other hand, if outside intensity of light is high, then the pupil contracts so that less light enters the eye.

It should be noted that the adjustment of the size of the pupil takes some time. For example, when we go from a bright light to a darkened cinema hall, at first we cannot see our surroundings clearly. After a short time our vision improves, and we can see the persons sitting around us. This is due to the fact that in bright sunlight the pupil of our eye is small. So, when we enter tht, cinema hall, very little light enters our eye and we cannot see properly.

After a short time, the pupil of our eye expands and becomes large. More light then enters our eye and we can see clearly. On the other hand, if we go from a dark room into bright sunlight or switch on a bright lamp, then we feel the glare in our eyes. This is due to the fact that in a dark room, the pupil of our eye is large. So, when we go out in bright sunlight, a large amount of light enters our eyes and we feel the glare. Gradually, the pupil of our eye contracts. Less light then enters our eye and we can see clearly.

Accommodation

A normal eye can see the distant objects as well as the nearby objects clearly. We shall now discuss how the eye is able to focus the objects lying at various distances. *An eye can focus the images of the distant objects as well as the nearby objects on its retina by changing the focal length of its lens.* The focal length of the eye-lens is changed by the action of ciliary muscles. The ciliary muscles can change the thickness of the eye-lens and hence its focal length which, in turn, changes the converging power of the eye-lens. Let us see how it happens.

When the eye is looking at a distant object (at infinity), the ciliary muscles of the eye are fully relaxed and the eye-lens is very thin (or less convex) in this position. Since the eye lens is very thin, its focal length is the maximum in this position and it can form the image of the distant object on the retina. When the eye is looking at a distant object, the eye is said to be 'unaccommodated' because it is the relaxed state of the eye. When the same eye has to see the nearby objects, the ciliary muscles become tense due to which

the eye-lens becomes thick (or more convex) and its focal length decreases. Due to short focal length, the converging power of eye-lens increases and it can focus the nearby objects on the retina. It has been focussed by the eye lens to form an image I on the retina. When the eye-lens becomes more convex to focus the nearby objects, the eye is said to be 'accommodated'. We can now say that: *The ability of an eye to focus the distant objects as well as the nearby objects on the retina by changing the focal length or converging power of its lens is called accommodation.* The maximum "accommodation" of a normal eye is reached when the object is at a distance of about 25 cm from the eye. After this the ciliary muscles cannot make the eye-lens bulge more. So, an object placed at a distance of less than 25 cm cannot be seen clearly by a normal eye because all the power of accommodation of the eye has already been used. Thus, *a normal eye has a power of accommodation which enables objects as far as infinity and as close as 25 cm to be focussed on the retina.*

RANGE OF VISION OF A NORMAL HUMAN EYE

The range of vision of a normal human eye is from infinity to about 25 centimetres. That is, a normal human eye can see the objects clearly which are lying anywhere between infinity to about 25 centimetres.

Far Point

The farthest point from the eye which can be seen clearly is known as the "far point" of the eye. The far-point of a normal human eye is at infinity.

Near Point

The nearest point upto which an eye can see the objects clearly is called the "near point" of the eye. The near-point of a normal human eye is at a distance of 25 centimetres from the eye.

Least Distance of Distinct Vision

The minimum distance at which an object must be placed so that a normal eye may see it clearly, is called the least distance of distinct vision (or least distance of clear vision). The least distance of distinct vision for a normal human eye is about 25 centimetres.

CONSTRUCTION OF A CAMERA

A simple camera consists of a light proof box having a convex lens L at one end and a light sensitive photographic film F at the other end. There is an arrangement for adjusting (or changing) the distance of the lens from the film. This is done by making the sides of the camera of folding canvas called bellows B. When the bellows are in folded position, then the distance between the lens and the film is the minimum. When we expand the bellows, the distance between the lens and the film gradually increases. As we shall see later on, this arrangement helps in focussing the distant objects as well as the nearby objects on the film of the camera. All of us have seen or handled a camera and we feel really very happy when we get photographed by it. Camera works on the refraction of light through a convex lens. When an object is placed in front of a camera, the camera lens forms a real, inverted and (usually) diminished image of the object on the photographic film inside it. We shall now describe the construction and working of a simple camera in detail.

There is a diaphragm D of variable aperture just behind the lens L. In the diaphragm we have a circular hole called aperture. The diameter of the aperture of diaphragm can be adjusted (increased or decreased) as desired. *The function of the diaphragm is to control the amount of light passing into the camera box through the lens.* As we shall study later on, the diaphragm of a camera is just like the iris of an eye because diaphragm controls the quantity of light entering a camera and iris controls the quantity of light entering an eye.

Behind the diaphragm is a *shutter S which is normally closed. Shutter is a device for opening and closing the aperture (or hole) of the diaphragm.* The shutter is opened only for a very short time to admit light into the camera while taking a photograph. Thus, the shutter opens or closes th entry of light rays to the film of the camera. A shutter is just like the door of a room whica can be opened or closed as required. The film can be rolled from outside by using the knob provided for this purpose. We shall now describe the working of this camera. A photographic camera is optically very similar to the human eye. It consists of a convex lens, a light-tight box, a light sensitive film to record an image and a shutter to let the light from the lens strike the film for a prescribed length of time. The lens forms a reduced real image of the object on the photographic film.

Working of a Camera

In order to take the photograph of an object, the camera is directed towards the object to be photographed with the lens facing the object. We should now focus the object so that a sharp and clear image of the object is obtained on the film. *The process of obtaining a sharp and clear image of an object is called focussing. The focussing in a camera is done by changing the distance between the lens and the film.* To focus the distant-object on the film, we have to decrease the distance between the camera lens and film by folding the bellows. On the other hand, to focus the near-by objects, we have to increase the distance between camera lens and film by expanding the bellows. Thus, in a camera, the distant objects as well as the near- by objects are focussed by changing the distance between the camera lens and the film.

Having focussed the object, we adjust the diameter of the aperture of the diaphragm according to the illumination or lighting of the object to be photographed. If the photograph is being taken in bright light, then the size of diaphragm aperture should be .small (so that excess light does not enter the camera) but if the photograph is taken in dim light then the aperture should be large (so that sufficient Sight reflected from the object may enter into the camera).

Finally, the shutter is opened for a fraction of a second and then closed again. When the shutter is opened, light rays coming from the object fall on the light-sensitive photographic film fixed in the camera and an image of the object is formed on the film. The exposed film is removed from the camera and developed to get the 'negative'. From negative, the actual photograph of the object is obtained.

Spherical Aberration

When we take a photograph with a camera, we want the image to be bright and sharp so that a clear photograph is produced. The brightness of image depends on the amount of light entering the camera. In other words, brightness of the image depends on the opening of the diaphragm or aperture. Now, when the diaphragm is opened wider to admit more light into camera, the image becomes less sharp. The image becomes less sharp because the edges of the camera lens may not focus the light to the same point as the centre part of the lens does. The inability of a convex lens to converge all the parallel rays of a wide beam of light to a single point is called spherical aberration. This defect of spherical aberration is corrected by using specially made meniscus lenses or by using a combination of lenses made of different types of glass instead of using a single lens. A combination of lenses made of different types of glass which do not produce spherical aberration are called "anastigmat lenses". In order to record the image properly on the film, the total light energy per unit area reaching the film (the exposure) must fall within certain limits. This is controlled by the shutter and the lens aperture. The shutter controls the time interval during which light enters the lens. This is usually adjustable in steps corresponding to factors of about two, often from 1s to l/l000s.

With a convex lens, the image distance decreases as the object distance increases. Hence in focussing the camera, the lens is moved closer to the film for a distant object and farther from the film for a nearby object. This is achieved by turning the lens in a threaded mount.

The intensity of light reaching the film is proportional to the area viewed by the camera lens and to the effective area of the lens. It is calculated that the intensity of light reaching the film with a particular lens is proportional to D^2/f^2.

The light gathering capability of a lens is commonly expressed in terms of the f-number of the lens. It is given by

$$f - \text{number} = \frac{\text{focal length}}{\text{diameter of the stop}} = \frac{f}{D}$$

Therefore, the light intensity reaching the film is inversely proportional to the square of the f-number.

Suppose the focal length of the lens is 11 cm and the diameter of the stop is 1 cm, then the f-number is 11 and D = f/11. The diameters of the stop of a camera are usually marked so that the exposure time is doubled between each position such as

$$\frac{f}{2}, \frac{f}{2.8}, \frac{f}{4}, \frac{f}{5.6}, \frac{f}{8}, \frac{f}{11}, \frac{f}{16} \text{ and so on.}$$

The larger numbers represent smaller apertures and exposures, and each step corresponds to a factor of two in intensity. The actual exposure is proportional to both to the aperture area and the time of exposure. Thus, f/4 and 1/500 s, f/5.6 and 1/250 s and f/8 and 1/125 s all correspond to the same exposure.

Camera Lenses

The simplest type of a camera lens is the biconvex lens. It suffers from every type of aberration. It was used in the earliest stages. With the advancement in technology, lens combinations are used. They are classified as normal, wide-angle, telephoto, zoom etc.

(i) The *normal lens* has a large aperture to reduce exposure time, an angular field of view of about 50° and produces an undistorted image.

(ii) A *telephoto lens* has a long focal length, gives a small angle of view and a large image of distant object.

(iii) A *wide-angle lens* is a lens of short focal length and gives a small image and a wide angle of view.

(iv) A *zoom lens* consists of a large number of individual lenses to decrease aberrations. The focal length of an ordinary camera may be increased up to six times without moving the image out of focus.

FIELD OF VIEW

Let us consider the situation. Light is passing through the aperture AB and reaching the eye E. The eye can view objects in the angle range PEQ. This is called the *Field of View*. If a concave lens is added in plane AB, light from a wider field P'Q' can reach the eye. We say *the field of view is enlarged.* On the other hand, if a convex lens is placed in plane

AB would narrow the field of view. In compound systems such as telescopes and microscopes, the situation is slightly different and we discuss about it at the appropriate places.

For a compound system the situation is slightly different. The cases of a telescope and a microscope respectively. In each case O is the centre of the objective, L_2 is the eye lens, and the image of O by L_2 is formed at centre E of the eye-ring. The outermost ray AE subtends angle θ_2 with the optic axis. So the eye placed at the exit pupil E receives a cone of rays of semi-angle θ_2. Now,

In the field of view is θ_1 given by

$$\theta_1 = \theta_2/m, \qquad ...(1)$$

where m is magnifying power of the telescope, and θ_2 is determined by the *eye lens*, aperture, the aperture of the objective has no role in this.

In a microscope the angle θ_2 is related to object height y (rather than (θ_1) and the magnification is defined by the ratio of θ_2 to y/D, where D is the least distance of distinct vision. We thus have

$$m = \frac{\theta_2}{y/D} \text{ or } y = \frac{\theta_2}{m}.D. \qquad ...(2)$$

The field of view y is again dependent on θ_2, the angle substended the eye lens at the exit pupil.

Need for a Multiple Lens Eyepiece

For a given magnification m, the field of view is proportional to the angle θ_2 subtended by the eyelens at the exit pupil (Eq. 1). For a single eye lens, the distance L_2E is close to focal length of L_2, so that θ_2 is approximately the ratio of the radial aperture to focal length of L_2. For a single lens this cannot be made large without serious aberrations. Considerable attention is therefore paid to constructing eyepieces with multiple lenses so as to produce a large field of view without much aberrations:

(a) *Enlarging the Field of View :* The field lens of the eyepiece effectively pushes the eye lens closer to the exit pupil, so as to enlarge the angle subtended.

(b) *Enabling us to Minimise Aberrations :* Firstly, the deviations of rays are divided over a larger number of surfaces, so that spherical aberration is minimised. Secondly, for a separated doublet chromatic aberration can be reduced to zero or considerably minimised without involving different materials for the lenses.

In Fig. 10.6 we show the general function of an eyepiece. In the two lens eyepieces the one nearer the eye is called *eye lens*, the other is calk *field lens*.

To appreciate the functions (a) and (b) cited above, let us take up the case of Ramsden's eyepiece the case. If $f'_1 = f_2' = 4.5$ cm, the focal length of the combination comes to 3.4 cm and distance of F' from the eye lens is 1.1 cm. This means that *far a given aperture* of eye lens the angle θ_2 obtained in the two-lens eyepiece is 3 times that obtainable by using a single lens of equivalent focal length (3.4 cm) and same aperture. Thus the field of view is considerably enlarged.

Also note that (for the same aperture) a single lens of f' = 3.4 cm will have more aberration than the eye lens of $f_2' = 4.5$ cm used in the two-lens eyepiece. The use of two separated lenses also reduces the chromatic and spherical aberrations as compared with a single lens.

For this reason Ramsden's eyepiece is called a *positive* eyepiece. One can use cross-hair in plane F, and the eyepiece would magnify the primary image and the cross-hair equally. With a graticule or a micrometer drive for the cross-hair, the primary image can thus be measured.

Representative Path of Rays

The location of the eye-ring centre E (exit pupil) differs from the focal point F'. For a telescope, the distance O (not shown) to F equals f', typically 50cm.

Hence, u = OH = 54.5 cm, and v = H'E = 4.9 cm.

Since H'F' = 4.5 cm, we get F'E = 0.4 cm.

Apertures of the Component Lenses

Aperture of L_1 ÷ aperture of L_2 is close to H'A' ÷ aperture of L_2, which equals H'E/L_2E.

H'E = 4.9 cm and $L_2E = 1/3\ (4.5) + 0.4$ cm + 1.9 cm. So the aperture ratio ≈ 5/2.

A single lens L_2' of same focal length f_E' will have to be placed at H'. Hence for same aperture, the θ_2' value would be $\theta_2 \times L_2E/H'E'$. Or, the enhancement of field of view (θ_2/θ_2') due to use of Ramsden's eyepiece is H'E'/L_2E. We have obtained this as ~ 5/2.

Gaussian Eyepiece

This is the same as the Ramsden's eyepiece, except that a glass plate inclined at 45° to the axis is placed between the two lenses as shown in Fig. 10.6 and a hole is provided on the side to admit light. This enables us to illuminate the back side of the cross-hair. This eyepiece is used to assist mechanical setting of a spectrometer, and also when faint images in a dark background are to be measured.

Kellner's Eyepiece

This eyepiece uses two plane-convex lenses with conditions.

$$f_1' = f_2' \quad \text{and} \quad t = f_2',$$

thus only slightly differing from Ramsden's eyepiece where t was taken somewhat smaller ($2f_2'/3$). In this eyepiece both the curved faces are put towards the incident light.

The eyepiece satisfies the condition of achromatism, but not the condition of minimum SA.

The relation of f_1', f_2' with f_E' is as follows:

$$f_E' = \frac{f_2' \times f_2'}{f_2' + f_2' - f_2'} = f_2' = f_1'.$$

The components thus have focal lengths (3/4) times smaller than those for the equivalent Ramsden's eyepiece. Hence spherical aberration is more.

The cardinal points have positions given by,

$$\alpha = f_E' \quad \alpha' = -f_E' \qquad \text{(check)}$$

Thus, H coincides with lens L_2 and H' coincides with lens L_1. When F, F' are plotted they coincide with L_1 and L_2 respectively.

That F' coincides with L_2 has a special significance. Since the exit pupil E is just to the right of F', L_2E is very small compared with f_E'. This factor would appear to give almost 15 to 20-fold gain of field of view if considerations of applied directly. But the exit pupil aperture is limited to ~ 3 mm diameter (due to eye pupil aperture ~ 2mm dia.) and when L_2 is so close it cannot usefully have the aperture of the normal eye lens. Hence, in practice the gain in field of view is ~5 for use with telescopes, which may reach –10 for use with microscopes.

Another characteristic of Kellner's eyepiece is that F coincides with lens L_1. The cross-hair are therefore engraved on the lens face itself. In fact, in most cases a fine scale is engraved on L_1 for measurement. This advantage has one accompanying trouble—any dust particles depositing on L_1 are seen in sharp focus!

Compound Microscope

While in a telescope the objective is a single lens or a contact doublet of large focal length, in a compound microscope the objective is a complex unit of very small f_0'. Our attention therefore shifts to the objective lens system. It will be useful to see why making f_0' small is the only option. The magnification is defined by

$$m = \tan\theta'/\tan\theta,$$

where, θ' is the angle subtended by the image (formed at infinity for convenience), and θ is the angle that would he subtended if the object were placed at the nearest distance of distinct vision, 25 cm.

PQ as an object of height y. From P a ray parallel to the axis and another through the focal point F_0 form the image at P'. For the final image at infinity, P' must lie on the focal plane F_E. The ray parallel to the axis must pass through f_E' and the other ray must emerge parallel to it. That gives angle θ'.

From the figure, we have

$$\tan\theta' = \frac{y'}{F_E'} \quad \text{and} \quad \frac{y'}{y} = \frac{x'}{f_0'}$$

with tan θ defined by $\frac{y}{25}$, we end up with

$$m = \frac{25x'}{f_0' f_E'} \qquad (x', f_0', f_E' \text{ in cm}) \qquad ...(3)$$

Now, the distance from second focus of objective (f_0') to first focus of eyepiece (f_E') is called the *optical length of a microscope*, and it has been standardised at 18 cm. Also, considerations of various aberrations and the fact that the exit pupil must not be smaller than pupil of the eye limit the lower practical value of f_E'. Therefore, all high power microscopes aim at securing lower value for f_0'.

High power microscope objectives require also a *large cone* of light to enter them, because, as we shall see later, the viewing of details depends upon this. With large cones of light, spherical aberration and coma cause serious troubles. To avoid spherical aberration we use a large number of lenses, including aplanatic lenses. To avoid chromatic aberrations we use achromatic lenses which are over-corrected to balance the chromatic errors of the first two aplanatic lenses also. Finally, to avoid coma we arrange the components so as to satisfy the Abbe sine condition. The designing of high power microscope objectives is thus a highly complex affair. Not only this, the use of these objectives involves accurate setting of the object and special methods of illumination. Therefore, a 'condensing system' for illuminating the object and a 'stage' for sample mounting are integral parts of all high power microscopes.

The smallest size of particles (or distance between two close points) which can be resolved (viewed separately) with a given microscope is given by

$$y_{min} = \frac{1.22\lambda}{\text{Numberical Aperture}}$$

where, λ is the wavelength of light used and *numerical aperture* is the product of the sine of the semi-angle of cone of light received by the object live and the refractive index of the medium in which the object is placed. But, we may see that use of smaller would enable us to observe smaller details. That is the secret of the *electron microscope.*

Ultra-Microscope

Let y_{min} be the smallest diameter of the object which can be resolved by a given microscope. Then if there are particles of diameter d *smaller* than y_{min} suspended in air or a liquid, this microscope is not able to see them ordinarily. Such particles are called *ultra-microscopic particles.*

If the medium is illuminated from a side so that no part of the *direct beam* enters the microscope objective, then the scattered rays from the particles only reach the microscope. The particles are then seen as bright stars in a *dark background.* This is the principle of the *ultra-micro scope.* The apparent size of the stars corresponds to y_{min}, not to d, s long as $d < y_{min}$. Hence only the presence of a particle is observed, not its size. All particles of size $< y_{min}$ appear as identical sized discs of diametery y_{min}. Note that not the microscope but the illuminating system is new. The system in used, for example, in Millikan's oil-drop experiment. The condensing system of a regular ultra microscope. The incident beam from the bottom enters through the annular opening in the diaphragm D, and the paraboloid condenser PP gives a hollow cone of light CC, such that all *direct* light goes outside the microscope objective. This is called *dark field illumination.* Only the scattered light from the object placed on plate S enters the microscope.

SIZE OF AN OBJECT

We often equate the *size of the object* to the linear dimension of the object, that is the height of the object. In practice, our feeling of the size of an object is actually derived from the size of its image cast on the retina. The size of the retinal image depends upon the visual angle, which is the angle subtended by the object at the eye. Therefore, it is known as the angular size.

A_1 and A_2 placed in front of the eye. The objects are of different sizes, but they subtend equal angles at the eye and their images formed on the retina are of the same size. If k is the distance between the retina and the lens,

$$\text{Thus} \qquad \frac{h_1}{k} = \theta_1 \quad h_1 \propto \theta_1 \qquad \text{or } h_1 = k\theta_1 \qquad \ldots(1)$$

Thus, the size of the image on the retina is proportional to the angle θ. Consequently, we bring objects close to our eyes to see them in more detail. This action makes the subtended angle and the retinal image as large as possible. As the object is brought closer to the eye (A_2 taken to the position A_3), the image on the retina becomes larger and larger on account of increase in the visual angle.

However, the object cannot be brought nearer to the eye beyond the least distance of distinct vision. The distance of distinct vision for a normal eye, NDDV, is 25 cm. If the object is brought nearer than this distance the eye cannot accommodate and in the process will have to strain itself.

An optical instrument forms the virtual image of the object, which acts as an object for the eye and subtends a much greater visual angle. A large image of a small object is thus formed on the retina. Thus, we see small objects which cannot be seen by the naked eye. The instruments used for increasing visual angle are called *magnifying instruments*.

THE SIMPLE MAGNIFIER

A converging lens can be used to form a virtual image that is larger and farther from the eye than the object itself. Then the object may be moved closer to the eye and the *angular size* of the image may be substantially larger than the angular size of the object held at 25 cm without the lens. A lens used in this way is called a *magnifier* or a *simple microscope*. The virtual image is most comfortable to view when it is placed at infinity, as the muscle of the eye is relaxed.

In object held at the near point (NDDV) where it subtends an angle θ at the eye. A magnifier is held in front of the eye forming an image at infinity and the angle θ' subtended at the magnifier.

$$\text{Angular magnification } M = \frac{\theta}{\theta'}$$

Note : Angular magnification is not the same as lateral magnification. Angular magnification is the ratio of the angular size of an image to the angular size of the corresponding object. On the other hand, lateral magnification is the ratio of the height of an image to the height of the corresponding object.

The difference can be understood this way. In the angular magnification is of about 2 times, since the object subtends an angle about twice larger than that without the lens. Hence the object looks about two times larger to the eye. However the lateral magnification is infinite, because the virtual image is at infinity. It does not mean that the object is infinitely large. When dealing with a magnifier M is useful but m is not.

Suppose, an object of length AB = h is placed at the near point A and viewed by the eye. Then the visual angle is α.

$$\tan\alpha = \frac{h}{D} \qquad ...(1)$$

We can calculate how much the lens magnifies. Suppose that a convex lens L of focal length f is used as a magnifying glass and the object is placed between O and F, such that the magnified erect image PQ is formed. If the observer's eye is close to the lens, then the distance OP is equal to the least distance of distinct vision. Here, the visual angle is given by

$$\tan\beta = \frac{PQ}{D} \qquad ...(2)$$

Then $$\frac{\tan\alpha}{\tan\beta} \cong \frac{\beta}{\alpha} = \frac{PQ}{h}$$

But $\frac{PQ}{D} = m = \frac{\upsilon}{u} = \frac{D}{u}$

Hence, angular magnification $M = \frac{\beta}{\alpha} = \frac{D}{u}$ ...(3)

We know that

$$\frac{1}{\upsilon} - \frac{1}{u} = \frac{1}{-D} - \frac{1}{-u} = \frac{1}{f}$$

(as per sign convention both D arid u are negative)

$$\therefore \quad -1 + \frac{D}{u} = \frac{D}{f} \text{ or } \frac{D}{u} = 1 + \frac{D}{f} \quad \text{...(4)}$$

Using the result (4) into eqn. (3), we obtain

$$M = 1 + \frac{D}{f} \quad \text{...(5)}$$

To see the image in the above case, the observer has to *accommodate* his eye. *Accommodation* means that the image is seen at the distance of most distinct vision, *i.e.* 25 cm.

When the object is placed at the focus of the lens, the image is formed at infinity. In this case, the image is seen *without accommodation.*

Then lateral magnification $m = \frac{PQ}{h} = \frac{\infty}{h} = \infty$

The angular magnification $M = \frac{\beta}{\alpha} = \frac{h/f}{h/D} = \frac{D}{f}$ ...(6)

From eqn. (6), it may seem that we can make the angular magnification as large as we like by decreasing the focal length. In fact, the aberrations of a simple lens set a limit to M of about 3× to 4×. Angular magnification may be made as great as 20× by correcting for aberrations. When greater magnification is required we use a compound microscope.

ENTRANCE PUPIL AND EXIT PUPIL

In D' is a diaphragm with opening A' B' in it and L is a lens in front of it. By backward drawing we obtain the points A, B in object space conjugate to A'B'. If an eye or any instrument is placed to the right of y, the width of the exit beam is limited by A'B', which is called the *exit pupil* for this highly simple single lens viewer. The width of the entrance beam is AB which is called the *entrance pupil.*

If, however, you introduce a diaphragm in plane A and its opening is *smaller* than AB, the situation changes. Imagine that the opening is (1/3) AB. Then the *entrance pupils* (1/3) AB and that limits the exit pupil to (1/3) A'B'

In this discussion it does not matter for us what is placed at the plane D or at plane D' – it may be a simple opening, or a convex lens, or a cave lens. All that we assume is that D is the *first* element on the entrance side, and D' is the *last* element on the exit side.

In the intervening space between D and D' there may be any number of elements—convex lenses, concave lenses, or both. All that we assume is that A'B' is the image of AB by all the intervening elements. In others words, the *exit pupil* is the image of the *entrance pupil.*

What we place in the opening AB is certainly of interest to us. If it is a simple opening we see objects in the projection of cone AOB. If there is a convex lens we see object in a smaller cone. If there is a plane mirror there, we see objects in a cone behind ourselves.

It may be noted that an eye placed at A' does not *see* the images of A, it only *receives the light* which passes through position A. If the eye extends over range A'B' it *receives* the light which enters the instrument over the range AB. What the eye really *sees* are objects to the left of AB, and in this business the elements placed in the aperture AB and A'B' do play a serious part.

In most situations the eye itself is the exit pupil. A diaphragm with an aperture is provided in the instrument in the plane where the image of the front lens (not of the object) is made by the intervening lenses. This aperture is slightly larger than the aperture of the human eye (~2mm); therefore, the human eye sets the limit at this exit end.

In a telescope the optimum ratio of exit pupil to entrance pupil is equal to 1/m, where m is the magnification. If a 40 cm diameter objective is used and magnification is 100, the exit pupil to use full objective aperture should be 4 mm. The eye would provide only about half the required pupil. Thus, there is waste of the outer half of the objective. For full use of the 40 cm diameter of the objective the magnification should be 200. For a camera or an epidiascope no such limitation exists.

STOPS AND PUPILS

When an optical instrument is used to study an object, light reflected from the object enters the instrument through its lenses. The light passing through the instrument is limited in cross-section either by the finite diameter of the lenses or diaphragms called *Stops*. The amount of light that reaches the image plane of the instrument depends on the brightness of the object and also on the dimensions of the stops located in the path of light. The pupil of the observer's eye should be placed at the exit pupil of a system in order to have a comfortable viewing of the image. Therefore, in an optical system stops and pupils are as much important as lenses or mirrors.

Aperture Stop : An aperture stop is a circular opening in an opaque screen. It limits the cross-section of the light that forms the image. Stops are used to restrict the light entering the instrument order to obtain a good image. A smaller aperture makes the final image less bright than before; but the size of the image remains the same. The *clear* diameter of the lens admitting the light into the instrument acts as an *aperture* while the peripheral rim acts as a *stop*.

Field Stop : It is also necessary to limit the size of the image if good images are required. For this purpose, we use a diaphragm known as the *Field Stop*. The field stop limits the field of view of the instrument. As an example, the film or plate holder acts as the field stop in a camera. As the field stop is made smaller, the field of view becomes narrower but the amount of light admitted to the image space remains constant. A field stop is often placed in the image plane.

Pupils : While a stop is a real object, a pupil is more conceptual. A pupil may be a real physical stop; or it may be an image of a stop. A *pupil* is defined as the cross-section of a bundle where light from all parts of the object passes through, completely mixed and without preferential spatial separation.

If we place a stop between the object and a lens, that stop will become the pupil. In this simple case of a single lens system, the stop is the *entrance pupil*. The lens not only forms the image of the object but also the image of the entrance pupil. The image of the entrance pupil is the *exit pupil*. Just as the object and image are conjugate, the entrance and exit pupils are conjugate. The location of the exit pupil depends upon the entrance pupil with respect to the lens. If the entrance pupil is kept at the left-hand focal plane of the lens, the exit pupil moves out to infinity on the right. If the stop is kept at the right-hand focal plane, then the exit pupil moves to infinity on the left. If the stop is kept within the focal length of the lens, the entrance pupil and exit pupil lie on the same side of the lens.

Combination of Lenses : The *entrance pupil* of a combination of lenses is the image of the aperture stop formed by all lenses preceding the stop. The image of the aperture stop, formed in the image space by all lenses following the stop is called the *exit pupil* or *eye ring*. The exit pupil may be located to the right of to the left of the entrance pupil, or even coincide with it. The exit pupil is the best position where the eye should be placed to collect as much light that passes through the objective as possible. If the eye *were* placed elsewhere the observer would find a narrower field of view.

A complex optical system consists of a combination of lenses and stops. There may be several diaphragms in addition to the aperture and field stops. The image of the field stop, formed by the lens preceding it, is called the *entrance window*. The image of the entrance window, as formed by the complete system, is called the *exit window*.

In visual instruments, the observer's eye is positioned at the centre of the exit pupil. The pupil of the eye varies from 2mm to 8 mm depending on the level of the illumination. An astronomical telescope mainly used at nights might have an exit pupil of atleast 8 mm.

REFRACTION AND TRANSLATION

A ray of light propagating through a cylindrically symmetric optical system undergoes two operations. At each surface boundary the direction of the ray changes due to *refraction*. In between the surfaces the height of the ray changes while the direction remains the same. This is known as *translation*. Therefore, to fully describe the ray propagation through an optical system, we make use of two operators, one for the refraction process and the other for the translation process. These two operators are known as *refraction matrix* and the *translation matrix* respectively.

There is a point to point correspondence between object space and the image space. Every point (x, y, z) in object space gets transformed into the point (x', y', z') in the image space. The transformation is linear in case of paraxial ray approximation and hence matrix operation can be used for the purpose. A ray is specified by its distance from the axis of the optical system and the angle that it makes with the axis. Let us consider a ray inclined to the axis and passing through the points A and B. The point A is at a distance x, from the axis. Also the ray is inclined at an angle α_1 at A with the z-axis, which is chosen as the axis of the optical system. If the ray makes an angle θ with the x-axis, then

$$\lambda_1 = \mu \cos \theta \qquad ...(1)$$

where α_1 is known as the optical direction cosine of the ray at A. The coordinates of the ray at A are specified by either (x_1, α_1) or (λ_1, x_1). Similarly, the coordinates of the ray at point B are given by either (x_2, α_2) or (λ_2, x_2).

Now let us consider a refracting curved surface of a cylindrically symmetric optical system (Fig. 3), the axis of symmetry being the z-axis.

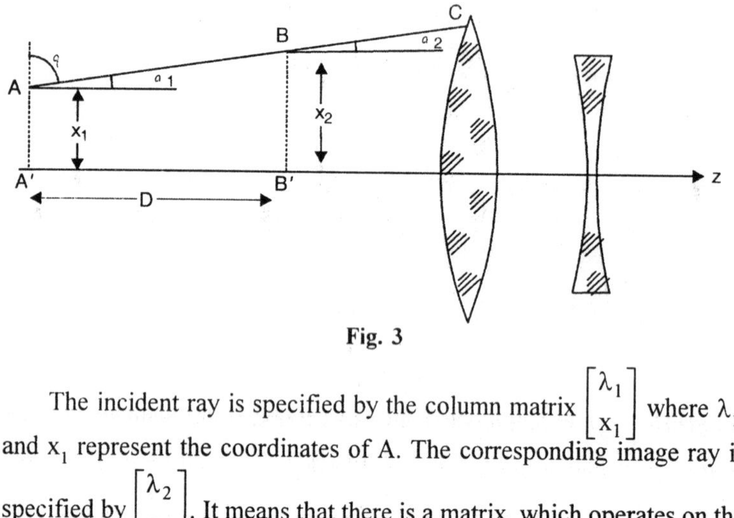

Fig. 3

The incident ray is specified by the column matrix $\begin{bmatrix} \lambda_1 \\ x_1 \end{bmatrix}$ where λ_1, and x_1 represent the coordinates of A. The corresponding image ray is specified by $\begin{bmatrix} \lambda_2 \\ x_2 \end{bmatrix}$. It means that there is a matrix, which operates on the column vector $\begin{bmatrix} \lambda_1 \\ x_1 \end{bmatrix}$ specifying the object to produce the column vector $\begin{bmatrix} \lambda_2 \\ x_2 \end{bmatrix}$ representing the image. We may now write that,

$$\begin{bmatrix} \lambda_2 \\ x_2 \end{bmatrix} = [T] \begin{bmatrix} \lambda_1 \\ x_1 \end{bmatrix}$$

or $$[R]\begin{bmatrix}\lambda_1\\x_1\end{bmatrix}=\begin{bmatrix}\lambda_2\\x_2\end{bmatrix} \quad ...(2)$$

The matrix [T] is called the *refraction matrix*. Note that the refraction matrix operating on the initial *coordinates gives the coordinates* of the refracted ray.

In the same way we can represent the effect of translation as

$$\begin{bmatrix}\lambda_2\\x_2\end{bmatrix}=[T]\begin{bmatrix}\lambda_1\\x_1\end{bmatrix}$$

or $$[T]\begin{bmatrix}\lambda_1\\x_1\end{bmatrix}=\begin{bmatrix}\lambda_2\\x_2\end{bmatrix} \quad ...(3)$$

The matrix [T] is called the *translation matrix*.

TRANSLATION MATRIX

Let us consider a paraxial ray traveling in a homogeneous medium of refractive index μ_1 which is at a distance of x_1 from the z-axis (axis of symmetry). Let A be a point on the ray. Let it be at a distance x_1 from the z-axis and be inclined at an angle α_1 with the z-axis. Thus, (x_1, α_1) are the coordinates of the ray at point A. Let (x_2, μ_2) be the coordinates of the ray at B through which the ray passes at a later instant. In traveling from A to B the ray undergoes *translation*. Since the medium is homogeneous, the ray travels in a straight line.

$$\alpha_1 = \alpha_2 \quad ...(1)$$

and $x_2 = x_1 + D, \tan\alpha_1 \approx x_1 + D\alpha_1$ ...(2)

Since the medium in which the light travels is the same, $\mu_1 = \mu_2$. It follows that,

$$\mu_1\alpha_1 = \mu_2\alpha_2 \quad ...(3)$$

or $$\lambda_1 = \lambda_2$$

where $$\lambda_1 = \mu_1\alpha_1$$

and $\lambda_2 = \mu_2\alpha_2$. ...(4)

Substituting the value of α, from equation (4) into equation (3), we get

$$x_2 = x_1 + D(\lambda_1/\mu_1)$$

which can be written as

$$x_2 = 1.\ x_1 + (D/\mu_1)\ \lambda_1 \qquad ...(5)$$

We can express λ_2 as follows:

$$\lambda = 1.\lambda_1 + 0.x_1 \qquad ...(6)$$

Equations (5) and (6) may be combined into the following matrix equation.

$$\begin{pmatrix} \lambda_2 \\ x_2 \end{pmatrix} = \begin{pmatrix} 1 & 0 \\ D/\mu_1 & 1 \end{pmatrix} \begin{pmatrix} \lambda_1 \\ x_1 \end{pmatrix} \qquad ...(7)$$

Thus, if a ray is initially given by a (2×1) matrix with elements x_1 and λ_1 then the effect of translation through a distance D in a homogeneous medium of refractive index μ_1 is characterized by a (2×2) matrix.

$$T = \begin{pmatrix} 1 & 0 \\ D/\mu_1 & 1 \end{pmatrix} \qquad ...(8)$$

The matrix is called *translation matrix*. It may be noted that

$$\det T = \begin{pmatrix} 1 & 0 \\ D/\mu_1 & 1 \end{pmatrix} = 1. \qquad ...(9)$$

SOLVED EXAMPLES

Example 1:

Find the position of the paraxial point of a lens in the from of a sphere of glass of radius 5 cm and refractive index 1.5 for the Fig. 7.6 is shown below.

Solution:

The system matrix can be obtained by multiplying the translation and refraction matrices in proper order.

$$S = \begin{pmatrix} 1 & 0 \\ u & 1 \end{pmatrix} \begin{pmatrix} 1 & -\dfrac{0.5}{5} \\ 0 & 1 \end{pmatrix} \begin{pmatrix} 1 & 0 \\ \dfrac{10}{1.5} & 0 \end{pmatrix} \begin{pmatrix} 1 & \dfrac{0.5}{5} \\ 0 & 1 \end{pmatrix}$$

$$= \begin{pmatrix} \dfrac{1}{3} & \dfrac{-2}{15} \\ \dfrac{u}{3} + \dfrac{2u}{3} & \dfrac{-2u}{15} + \dfrac{1}{3} \end{pmatrix}$$

$$\begin{pmatrix} \lambda_2 \\ x_2 \end{pmatrix} = \begin{pmatrix} \frac{1}{3} & \frac{-2}{15} \\ \frac{u}{3}+\frac{2u}{3} & \frac{-2u}{15}+\frac{1}{3} \end{pmatrix} \begin{pmatrix} \lambda_1 \\ x_1 \end{pmatrix}$$

or $$x_2 = \left(\frac{u}{3}+\frac{2u}{3}\right)\lambda_1 + \left(\frac{-2u}{15}+\frac{1}{3}\right)x_1$$

Here $$x_2 = 0 \text{ and } \lambda_2 = 0$$

$$\left(\frac{-2u}{15}+\frac{1}{3}\right)x_1 = 0.$$

Example 2:

Find the focal length, and unit planes of Ramsden's eye-piece.

Solution:

Ramsden's eye piece consists of two plano convex lenses each of focal length f separated by a distance of $\frac{2f}{3}$. The Ramsden's eye piece.

The system matrix for the lens $L_1 = \begin{pmatrix} 1 & \frac{-1}{f} \\ 0 & 1 \end{pmatrix}$

The translation matrix from L_1 to $L_2 = \begin{pmatrix} 1 & 0 \\ \frac{2f}{3} & 1 \end{pmatrix}$

The system matrix for the second lens $= \begin{pmatrix} 1 & \frac{-1}{f} \\ 0 & 1 \end{pmatrix}$

The system matrix of Ramsden's eye piece

$$S = \begin{pmatrix} 1 & \frac{-1}{f} \\ 0 & 1 \end{pmatrix} \begin{pmatrix} 1 & 0 \\ \frac{2f}{3} & 1 \end{pmatrix} \begin{pmatrix} 1 & \frac{-1}{f} \\ 0 & 1 \end{pmatrix} = \begin{pmatrix} \frac{1}{3} & \frac{-4}{3f} \\ \frac{2f}{3} & \frac{1}{3} \end{pmatrix}$$

Comparing this with the general system matrix, Gaussian constants are

$$a = \frac{4}{3f};\ b = \frac{1}{3};$$

$$c = \frac{1}{3};\ d = \frac{-2f}{3}$$

The combined focal length of the system F is given by

$$F = \frac{1}{a} = \frac{3f}{4}$$

The distance of the unit planes

$$\alpha = \frac{1-b}{a} = \frac{1-(1/3)}{(4/3f)} = \frac{2/3}{4/3}xf = \frac{1}{2}f$$

$$\beta = \frac{c-1}{a} = \frac{(1/3)-1}{4/3} = \frac{-1}{2}f$$

Nodal planes coincide with unit planes.

Example 3:

In a telescope with a single lens eyepiece of $f = 6$ cm the objective is 106 cm from the eyepiece. Deduce the position of the exit pupil magnification of the telescope, and size of entrance pupil if exit pupil is taken as 2 mm diameter. If diameter of the eye lens is 3 cm, deduce the semiangle of the field of view.

Solution:

Exit pupil is at the image of objective by the eye lens.

Given u = – 106 cm, f = 6 cm,

we get the exit pupil position

$$v = \frac{106 \times 6}{100} \text{ cm} = 6.4 \text{ cm from the eye lens.}$$

For a telescope, magnification = angle subtended by eye lens at exit pupil ÷ angle subtended by eye lens at the objective. Hence,

$$m = 106 \div 6. = 16.7$$

Size of entrance pupil ÷ size of exit pupil = magnification. Hence diameter of entrance pupil is

$$d \text{ (entrance)} = 2 \text{ mm} \times 16.7 = 3.3 \text{ cm.}$$

From semi-angle θ_1 is angle subtended by eye lens at the objective.

$$\theta_1 = \frac{1.5}{106} = .0142 \text{ radian}$$

Example 4:

A Huygens' eyepiece having focal ratio 2.5 and satisfying achromatism is to be designed. Deduce focal lengths of the components *and the cardinal points in terms of f_E'. In what respects does it differ from the Huygens' eyepiece with focal ratio 3?*

Solution:

$$f_E' = \frac{(2.5f_2')(f_2')}{2.5f_2'+f_2'-\frac{1}{2}(2.5f_2'+f_2')} = \frac{2.5f_2'}{1.75} = \frac{10}{7}f_2'$$

$$\therefore \quad f_2' = \frac{7}{10}f_E', \quad f_1' = \frac{7}{4}f_E', \; t = \frac{49}{40}f_E'$$

Note that, though *ratio* f_2'/f_1' is smaller than in standard Huygens' eyepiece, actual f_2' is slightly larger (hence smaller spherical aberration).

Next one may see that

$$L_1H = \alpha = \frac{7}{10}f_E',$$

$$L_2H' = \alpha' = -\frac{7}{10}f_E'$$

$$L_2F' = \alpha' = +f_E' = \frac{3}{10}f_E';$$

$$L_1F = \alpha - f_E' = \frac{3}{4}f_E'$$

The first focal plane F is still to the right of L_1. The eyepiece is thus negative. The quantity f_E'/L_2F', which determines enlargement of the field, is 10/3. This is larger than in the normal Huygens' eyepiece.

Example 5:

If aperture of the eye is taken as 2 mm and least distance of distinct vision is 25 cm, calculate the 'numerical aperture' of the unaided eye. The best numerical aperture obtained in a microscope approaches a value 2. How much should the normal magnification of such microscope be?

Solution:

Semiangle = 0.1 cm/25 cm = 0.004 radian. For air n = 1. Hence.

N.A. of the eye = 0.004 radian

Considering 1.22 λ term common between eye and a microscope, the detail resolved by the objective O the microscope is smaller by factor

$$\frac{\text{N.A. of microscope}}{\text{N.A. of the eye}} = \frac{2}{.004} = 500$$

So magnification should be > 500 fold if the microscope is to use full resolution. 'Normal' magnification = 500.

Example 6:

Let the focal length of an eyepiece be 2.5 cm. For a telescope of magnification 100, deduce the focal length of the objective and useful diameter of the objective, given d_{eye} = 2 mm.

Solution:

Given $m = \dfrac{f_0'}{f_E'} = 100$ and $f_E' = 2.5$ cm

$\therefore \qquad f_0' = 250$ cm

For optimum use, $\dfrac{d_0}{d_E} = m$;

and $d_E = 2$ mm (given)

$d_0 = 20$ cm.

Example 7:

In the case of Example 6.1 add a lens of f' = 12 cm at the position of the primary image, giving it a 'suitable' aperture to use the full aperture of the eyelens. Calculate (i) aperture of this field lens, (ii) new semi-angle of the field of view.

Solution:

With O as centre of the objective, OBE is the path of marginal rays when L_2 alone is present. With L_1 added, the path of marginal rays becomes OABE'. Here lens L_1 is the field lens added, making $L_1 + L_2$ the two-lens eyepiece. We have

$OL_1 = 100$ cm, $L_1L_2 = 6$ cm

whence we deduce (check them);

$L_1P = 13.6$ cm

$\Rightarrow \qquad L_2P = 7.6$ cm

aperture of $\quad L_1 = 3 \times \dfrac{13.6}{7.6} = 5.3$cm

now $\quad \theta_1 = \dfrac{1}{2} \text{x} \dfrac{5.3}{100} = .0265$ radian.

Thus the field of view increases to about 2-fold. You may note that lens L1 here does not play a role in image formation (because its location coincides with the primary image); yet it enhances the field of view. The eye ring shifts from E to E', so θ_2 also increases by the same factor as θ_1, keeping the magnification unchanged.

Example 8:

Find the focal length, unit planes of Huygens's eye piece.

Solution:

Huygens's eye piece consists of two Plano convex lenses each of focal lengths 3f and f separated by a distance of 2f. The Huygens's eye piece.

The system matrix for the lens $L_1 = \begin{pmatrix} 1 & \frac{-1}{3f} \\ 0 & 1 \end{pmatrix}$

The translation matrix from L_1 to $L_2 = \begin{pmatrix} 1 & 0 \\ 2f & 1 \end{pmatrix}$

The system matrix for the second lens $= \begin{pmatrix} 1 & \frac{-1}{f} \\ 0 & 1 \end{pmatrix}$

The system matrix of Ramsden's eye piece

$$S = \begin{pmatrix} 1 & \frac{-1}{f} \\ 0 & 1 \end{pmatrix}\begin{pmatrix} 1 & 0 \\ 2f & 1 \end{pmatrix}\begin{pmatrix} 1 & \frac{-1}{3f} \\ 0 & 1 \end{pmatrix} = \begin{pmatrix} 1 & \frac{-2}{3f} \\ 2f & \frac{1}{3} \end{pmatrix}$$

Comparing this with the general system matrix, Gaussian constants are

$$a = \frac{2}{3f};\ b = -1;$$

$$c = \frac{1}{3};\ d = -2f$$

The combined focal length of the system F is given by

$$F = \frac{1}{a} = \frac{3f}{2}$$

The distances of the unit planes

$$\alpha = \frac{1-b}{a} = \frac{1+1}{2} x3f = 3f$$

$$\beta = \frac{c-1}{a} = \frac{(1/3)-1}{2} x3f = f$$

Nodal planes coincide with unit planes.

Example 9:

Obtain the system matrix for a thin lens placed in air and made of refractive index 1.5 and radii of curvature 50 cm each.

Solution:

System matrix for a thick lens is given by

$$S = \begin{pmatrix} 1 & -P_1 - P_2 \\ 0 & 1 \end{pmatrix}$$

The powers of two surfaces are equal. Hence

$$P_1 = P_2 = \left(\frac{\mu - 1}{R_1}\right)$$

$$= \left(\frac{\mu - 1}{|R_2|}\right) = \frac{1.5-1}{50} = 0.01$$

$$\therefore \quad S = \begin{pmatrix} 1 & -0.01-0.01 \\ 0 & 1 \end{pmatrix}$$

$$= \begin{pmatrix} 1 & -0.02 \\ 0 & 1 \end{pmatrix}$$

Further $\qquad f = \dfrac{1}{0.02} = 50$ cm.

Example 10:

Two thin converging lenses of powers 5 and 4 diopters are placed co-axially 10 cm apart. Find the focal length of the combination.

Solution:

Power of a lens $P = \dfrac{1}{f}$

$$f_1 = \frac{1}{P_1} = \frac{1}{4} = 0.25 \text{ m} = 25 \text{ cm}$$

$$f_2 = \frac{1}{P_2} = \frac{1}{5} = 0.20 \text{ m} = 20 \text{ cm}$$

$$\frac{1}{F} = \frac{1}{f_1} + \frac{1}{f_2} - \frac{t}{f_1 f_2}$$

$$F = \frac{f_1 f_2}{f_1 + f_2 - t}$$

$$= \frac{25 \times 20}{25 + 20 - 10} = 14.3 \text{ cm.}$$

Example 11:

A thick bi-convex lens having radii of curvature + 6 cm and --6 cm is made up of a material of refractive index 1.6. If the thickness of the lens is 2 cm. Find :

(a) System matrix

(b) Focal length

(c) Unit planes.

Solution:

Radii of curvature of lens $|R_1| = |R_2| = 6$ cm

Refracting powers of the surfaces

$$P_1 = P_2 = \left(\frac{\mu - 1}{R}\right) = \frac{1.6 - 1}{6} = 0.1 \text{ cm}^{-1}$$

System matrix of the lens

$$S = \begin{pmatrix} 1 - P_2 t/\mu & -P_1 - P_2(1 - P_1 t/\mu) \\ \dfrac{t}{\mu} & 1 - P_1 t/\mu \end{pmatrix}$$

$$= \begin{pmatrix} 1 - \dfrac{0.1 \times 2}{1.6} & -0.1 - 0.1(1 - \dfrac{2}{1.6} \times 0.1) \\ \dfrac{2}{1.6} & 1 - \dfrac{0.1 \times 2}{1.6} \end{pmatrix}$$

$$= \begin{pmatrix} 0.875 & -0.1875 \\ 1.25 & 0.875 \end{pmatrix}$$

Comparing with the general system matrix,

$a = 0.1875;\ b = 0.875;$

$c = 0.875;\ d = -\ 1.25$

Focal length of the lens

$$f = \frac{1}{a} = \frac{1}{0.1875} = 53 \text{ cm.}$$

Distances of unit planes,

$$\alpha = \frac{1-b}{a} = \frac{1-0.875}{0.1875} = 0.67 \text{ cm}$$

$$\beta = \frac{c-1}{a} = \frac{0.875-1}{0.1875} = -\ 0.67 \text{ cm}$$

Thus, first unit plane is at a distance of 0.67 cm to the right of the first refracting surface, where as the second unit plane is at a distance of 0.67 cm to the left of the second refracting surface.

Example 12:

The focal lengths of two thin convex lenses are 20 cm and 10 cm. They are at a distance of 10 cm on the axis of symmetry. If a point object of length 1 cm is placed at a distance of 20 cm from the first lens, find the position of the image and magnification.

Solution:

Let the image be formed at a distance of v cm from the second lens. The system matrix can be obtained in the following way.

The translation of a matrix between object and lens L_1 is given by

$$\begin{pmatrix} 1 & 0 \\ d & 1 \end{pmatrix} = \begin{pmatrix} 1 & 0 \\ 20 & 1 \end{pmatrix}$$

The refraction matrix for lens L_1 is given by

$$\begin{pmatrix} 1 & \dfrac{-1}{f_1} \\ 0 & 1 \end{pmatrix} = \begin{pmatrix} 1 & -\dfrac{1}{20} \\ 0 & 1 \end{pmatrix}$$

The translation matrix between the lens L_1 and L_2 is given by

$$\begin{pmatrix} 1 & 0 \\ d & 1 \end{pmatrix} = \begin{pmatrix} 1 & 0 \\ 10 & 1 \end{pmatrix}$$

The refraction matrix lens L_2 is given by

$$\begin{pmatrix} 1 & \frac{-1}{f_2} \\ 0 & 1 \end{pmatrix} = \begin{pmatrix} 1 & -\frac{1}{10} \\ 0 & 1 \end{pmatrix}$$

The translation matrix between L_2 and image is given by

$$\begin{pmatrix} 1 & 0 \\ d & 1 \end{pmatrix} = \begin{pmatrix} 1 & 0 \\ v & 1 \end{pmatrix}$$

Now the system matrix is given by

$$S = \begin{pmatrix} 1 & 0 \\ v & 1 \end{pmatrix}\begin{pmatrix} 1 & -\frac{1}{10} \\ 0 & 1 \end{pmatrix}\begin{pmatrix} 1 & 0 \\ 10 & 1 \end{pmatrix}\begin{pmatrix} 1 & -\frac{1}{20} \\ 0 & 1 \end{pmatrix}\begin{pmatrix} 1 & 0 \\ 20 & 1 \end{pmatrix}$$

$$= \begin{pmatrix} -2 & \frac{-1}{10} \\ -2v+20 & \frac{-v}{10}+\frac{1}{2} \end{pmatrix}$$

For image plane we must have

$$-2u + 20 = 0$$

or $$U = 10 \text{ m.}$$

Hence the image will be formed at a distance 10 cm from the second lens.

The magnification m is given by

$$m = \frac{-v}{10}+\frac{1}{2} = \frac{-1}{2}.$$

EXERCISES

1. Discuss the function of an eye-ring (exit pupil) in any viewing instrument.
2. Give the construction and working of a Ramsden eyepiece. How are chromatic and spherical aberrations minimized in this eyepiece? Indicate in a diagram the position of the cardinal points. Make the necessary calculations.
3. Explain with the help of a neat diagram the construction and working of a Huygens eyepiece and clearly indicate the positions

of its cardinal points. Why is it referred to as a theoretically perfect but a negative eye-piece?

4. What is an eye-piece and what is its advantage over a single lens? Give with a neat sketch the theory of Ramsden eyepiece. What types of eyepieces would you recommend for:

   (i) low power microscope

   (ii) spectrometer telescope?

5. In a microscope, the first image is formed 30.0 cm behind the objective, and an eyepiece of fE' = 5.0 cm is set for viewing with a relaxed eye, Deduce the position of the eye-ring relative to F' of the eye-piece.

6. Why is a double-lens eye-piece better than a singe lens eyepiece of equivalent focal length? Explain.

7. A telescope objective is rendered achromatic by using a contact doublet, but an eyepiece to be rendered achromatic uses separated lenses. Discuss.

8. Deduce the composition of a feamsden's eyepiece of equivalent focal length 4.2 cm. Derive and plot its cardinal points. Show the path of a pencil of rays through it, covering the full aperture of the eye lens.

10. Distinguish between a *'positive'* and a *'negative'* eye-piece. If a crosswire is to be used with a Huygens eyepiece, show where it should be located. What are the limitations, if any, of such use?

11. Deduce f_1', f_2', t, eye-ring position and aperture ratio of an eyepiece to be used with a telescope whose objective has focal length 60.0 cm, given f_E' = 4.5 cm; F at 1.0 cm in front of the field lens and F' at 0.5 cm behind the eye lens.

12. What do you understand by visual angle and angular magnification? Give the construction and working of a simple microscope. Calculate its magnifying power.

13. Describe a simple photographic camera. What do you understand by f-number? Give the construction of a telephoto lens and explain its working.

14. What are the functions of a field lens used in an eyepiece? Give the construction of a Huygens eyepiece and calculate the positions of the cardinal points.

16. What is an eye-piece and what is its advantage over a single lens?
17. Give the construction and working of a Ramsden eyepiece. How are chromatic and spherical aberrations minimized in this eyepiece?
18. Explain the construction of a Huygens eyepiece. Why cannot a cross-wire be used with it?
19. Give the name and construction of the eyepiece, which satisfies the condition for achromatism.
20. Describe and point out the respective merits of Ramsden and Huygens eyepieces.
21. Explain the principle of an oil-immersion objective.
22. Explain what do you understand by spherical and chromatic aberrations. Describe how these are minimized in a Huygens eyepiece.
23. Aperture of the eye is ~ 2 mm in diameter and effective A for vision is $- 5 \times 10^{-5}$ cm. Deduce the smallest particle size which can be resolved by an appropriate microscope of magnifying power 60. (Take least distance of distinct vision as 25 cm). What is to be 'appropriate' about the microscope?
24. a pencil of rays corresponding to an intermediate image point P near the edge of the field of view is traced. Draw the corresponding diagram for a point at the centre of the field of view (*i.e.*, P coinciding with F).
25. Show that in a telescope the field of view depends on magnification m and the angle subtended by the eye lens at the eye, and is independent of aperture of the objective. What is then the need for having a large aperture of the objective?
26. What is System matrix? Obtain it in the case of a system of two thin lenses separated by a distance and hence derive the formula for its focal length.
27. Explain how translation and refraction matrices are formed and hence build up the system matrix for a thick lens.
28. The focal length of each lens of Ramsden's eye-piece is 4 cm by matrix formulation. Find the focal length and determine the positions of the cardinal points.

29. The radius of curvature of the surfaces of a double convex lens are $R_1 = R_2 = 50$ cm. The refractive index of the material of the lens is 1.5. Find the optical power of the lens.
30. Using matrix method obtain the expression for equivalent focal length and position of cardinal points of Huygens's eye piece?
31. Formulate the equations for the image and magnification of an optical system using the matrix method?
32. What are unit planes and nodal planes of a kens system? Find the positions of the unit planes in a thick lens and hence deduce the expression for its focal length.
33. Obtain the system matrix for the thick lens and drive the lens formula for thin lens.
34. Show that if the distances are measured from the unit planes, the focal length of the system is given by the reciprocal of the element of the system matrix.
35. Using matrix method obtain the expression for equivalent focal length and position of cardinal points of Ramsden's eye piece?
36. What is the advantage of using matrix method in paraxial optics? Obtain the *system matrix* for a general optical system.
37. Determine the matrices that will represent the effect of translation and refraction in a medium of refractive index μ.

4

ROTATING MIRROR METHOD BY FOUCAULT'S

INTRODUCTION

The Foucault's arrangement for measuring the speed of light. Light from an illuminated slit A, whose length is perpendicular to the plane of paper, is reflected from 45° glass plate P and rendered very nearly parallel by lens L. It is then reflected by a plane mirror R_1 and is focussed on concave mirror C placed at a distance D, which is a *few metres*, R can rotate about an axis perpendicular to the plane of paper. This axis of R passes through the centre of curvature of the mirror C; hence, the rays retrace their path on reflection at C. If R is not rotating, the returning rays form an image B. If R rotates by angle a in the time taken by light in travelling the distance 2D, the return beam is deflected by angle 2α with respect to the onward path:. Hence the image is formed at B_1; the displacement BB_1 = s is measured by means of an eyepiece. It equals f. 2α where f is the focal length of lens L.

If mirror R makes n rotations/sec, we have

$$s = f.2\alpha = 2f.\ 2\pi n \times 2D/c$$

This gives, $$c = \frac{8\pi nf.}{s} D \qquad ...(1)$$

With D~20 metres, the shift s obtained by Foucault was about 1 mm, the measurement of which could be accurate to about 1%. Increase of D would cause lowering of intensity; for, if aperture of mirror C is < d, then during the rotation of R the mirror C receives light for only a fraction $d/2\pi D$ of the time in each rotation.

SPEED OF LIGHT IN A TRANSPARENT LIQUID

In 1862, Foucault designed an apparatus for measurement of velocity of light. It requires much shorter distance than Fizeau's method and it is a modification of Fizeau's method. Foucault's arrangement for this purpose is an adaptation. Only the chief rays of beams are shown here. The change is that two concave mirrors C_1 and C_2 are used, equidistant from the rotating plane mirror R. In the path of light to one of the mirrors a tube T containing the experimental liquid is introduced. Auxiliary lenses L_1 and L_2 are needed to make the beam pass through the tube T as a parallel beam and to focus the emergent beam on C_2. When the rotating mirror is in position L_1 it sends a flash of light on C_1, which returns at the instant when mirror is in position R_1'.

The return beam gets focussed at B_1, where BB_1 is the displacement. When the mirror is in position R_2, it sends a flash of light on C_3, which returns at the Instant when mirror is in position R_2'. The return beam therefore, focuses at another point B_2, since time of flight is changed.

Let $RC_1 = RC_2 = D$. In path RC_2 the tube T, containing the liquid copies almost the entire length. Therefore, if c' is the speed of light liquid, we have

$$\frac{BB_2}{BB_1} \equiv \frac{s'}{s} = \frac{c}{c'} \qquad \text{...(2)}$$

The measurement, when first made by Foucault (for water), was not very accurate. Yet it was of historic significance. He found s' > s, so that c' was less than c. *Speed of light in water was found to be less than its speed in air.* The result gave a death blow to the corpuscular theory of light which predicated c' > c. Further, the ratio c/c' was found to be equal to retractive index of the liquid, in *quantitative* support of the wave theory. In other words Foucault used a rotating mirror instead of a toothed wheel. Light from a strong source S falls on an achromatic lens L, after passing through the glass plate P. The light after passing through the lens L will converge at the point I.

If a mirror M_1 is placed at A, light after reflection from the surface of the mirror M_1, converges at the pole of concave mirror C, whose distance from A is adjusted equal to its radius of curvature. Light is reflected back from C along its original path and finally the image is formed on the source S. As there is a half silvered glass plate P inclined at an angle of 45° to the axis of the lens, the image is formed at B_1, which can be viewed with the help of a micrometer eyepiece.

Suppose, the plane mirror M_1, is rotated rapidly at a uniform speed about an axis passing through A, the rays after reflecting from the concave mirror C, find the plane mirror displaced by an angle θ to a new position M_2. The image is now observed at B_2. The displacement B_1B_2 of the image is measured and the velocity of light is calculated.

Theory : Consider a point E on the concave mirror from which light is reflected back to the rotating mirror. When the plane mirror is at the position M_1, the image of S is at I such that AE = AI = α, where a is the radius of curvature of the concave mirror. When the plane mirror is rotating with a uniform speed and is at a position M_2, the image of S will be formed at I and the light appears to diverge from I_1 instead of I.

If the mirror turned through an angle θ, the reflected ray is turned through an angle 2θ.

$$I\,I_1 = \theta \times 2\theta$$

As S and S_1 are conjugate points with respect to I and I_1, for the lens L

$$\frac{SS_1}{d} = \frac{II_1}{(a+b)}$$

$$S_1S = \frac{a \times 2\theta . d}{(a+b)}$$

Suppose $\quad S_1S = B_1B_2 = x$

Then $\quad S_1S = \dfrac{a \times 2\theta . d}{(a+b)}$

and $\quad \theta = \dfrac{x(a+b)}{2ad} \qquad ...(3)$

If the plane mirror makes n rotations per second, then the time taken by the plane mirror to rotate through an angle θ is,

$$t = \frac{\theta}{2\pi\ n}$$

During this time t, light travels from A to E and back again to A.

$\therefore \quad t = \dfrac{2\alpha}{c}$ where c is the velocity of light.

$$\frac{2\alpha}{c} = \frac{\theta}{2\pi n};\ \theta = \frac{4\pi n a}{c} \qquad ...(4)$$

From eqn. (7) and (8), we obtain

$$\frac{x(a+b)}{2ad} = \frac{4\pi n a}{c}$$

$$\therefore \qquad c = \frac{8\pi n a^2 d}{x(a+b)} \qquad ...(5)$$

As n, a, d, x and b are measurable quantities, c can be calculated.

The displacement of the image in the Foucault's experiment was only 0.70 mm and could be measured with the help of micrometer eyepiece. The value of c found by Foucault was 2.96×10^8 m/s.

The main disadvantage of the Foucault's method is that the image obtained is not very bright due to reflection and refraction of light at various surfaces.

Foucault observed the velocity of light in water by placing a long cylindrical pipe of water between the concave mirror and the plane mirror. He found that the displacement x of the image when water was used, was greater for the same speed of rotation of the mirror than when air was used.

Therefore, according to the relation $c = \frac{8\pi n a^2 d}{x(a+b)}$, the value of c in water should be less than the value of c in air.

Newton's corpuscular theory of light predicted that the velocity of light in water should be more than that in air. The result of Foucault gave a death blow to the corpuscular theory of light. The wave theory of light predicted that the velocity of light in water should be less than in air. Thus, Foucault's experiment justified the validity of the wave theory of light.

DISPERSION

A beam of white light, when it passes through a prism is split up into its constituent colours. Different colours suffer different amounts of deviation. The spread of colours is called *dispersion of light*. The image thus, formed on a screen is called a *spectrum*. The medium, which produces dispersion, is called a *dispersive medium*.

The spectrum consists of visible and invisible regions. In the visible region the order of the colours is from violet to red. The principal colours are given by the word *VIBGYOR* (Violet, Indigo, Blue, Green, Yellow,

Orange and Rod). The deviation produced for the violet rays of light is maximum and for red rays of light it is minimum. Fig. 8.1 represents the dispersion of a white ray of light by a prism in the visible region. The region of the spectrum of wavelengths shorter than violet is called *ultraviolet* and the region of wavelengths longer than red is called *infrared*. In the present chapter, the discussion relates only to the visible region of the spectrum.

It is evident the refractive index for the material of a prism (or a lens) is different for different wavelengths (or colours). The deviation and hence the refractive index is more for blue rays of light than the corresponding values for red rays of light. The deviation and the refractive index of the yellow constituent are taken as the mean values different for different wavelengths (or colours). The deviation and hence, the refractive index is more for blue rays of light than the corresponding values for red rays of light. The deviation and the refractive index of the yellow constituent are taken as the mean values.

REFRACTION THROUGH A PRISM

The refractive index of the material of prism is given by

$$\mu = \frac{\sin \frac{A+\delta_m}{2}}{\sin \frac{A}{2}} \qquad ...(1)$$

where A the angle of the prism and δ_m is the angle of minimum deviation. When the angle of the prism is very small, the prism is said to be a *thin prism*. For a thin prism, the values of the angles A and δ are so small that the sines of the angles may be taken equal to the angles. Therefore, eqn. (1) becomes

$$\mu = \frac{A+\delta_m}{A}$$

or

$$\delta_m = (\mu - 1)A \qquad ...(2)$$

ANGULAR DISPERSION

The refraction and dispersion does not bear a simple relationship to one another. It is known that some glasses have a high index of refraction and little dispersion while others have low index of refraction and high dispersion. Therefore, to characterize a material, it is necessary to use more than one refractive index. In practice three colours are chosen; one in the middle and two at the two extremes of the visible spectrum. Thus, the yellow colour is taken as the mean colour and violet and red colours are considered at the two extremes. The refractive indices corresponding to these colours are indicated by μ, μ_V and μ_R. The deviation 5 corresponding to yellow colour is taken as *mean deviation.* The angles of deviation δ_V, δ and δ_R produced in the violet, mean yellow and red rays of light.

The deviation δ_v, δ and δ_R can be written as

$$\delta_v = (\mu_v - 1)A \quad ...(1)$$

$$\delta_R = (\mu_R - 1)A \quad ...(2)$$

$$\delta_R(\mu_R - 1)A. \quad ...(3)$$

The total angle through which the spectrum is spread is called as the *angular dispersion.* It is measured by the difference in deviation between the two extreme colours of the spectrum. Therefore, the angular dispersion is given by

$$\theta = \delta_v - \delta_R = (\mu_v - \mu_R) A. \quad ...(4)$$

Thus, the angular dispersion depends on the nature of the material of the prism and upon the angle of the prism.

DISPERSIVE POWER

Dispersive power indicates the ability of the material of the prism to disperse the light rays. It is defined as the ratio of the angular dispersion to the deviation of the mean ray.

$$\text{Dispersive power, } \omega = \frac{\delta_V - \delta_R}{\delta} \quad ...(1)$$

Using the set we can rewrite eqn. (1) as

$$\omega = \frac{\mu_V - \mu_R}{\mu - 1} \quad ...(2)$$

This is the expression for *dispersive power* used mainly in *geometric optics* and stated in terms of mean μ and refractive indices for violet and red light. The difference $(\mu_v - \mu_R)$ is known as the *mean dispersion.* It is seen that the dispersive power is independent of the angle of the prism and the angle of incidence.

Note : There is no universal choice about colours chosen for defining the dispersive power. In case of viewing with the aid of eyes, the range violet to red is more useful, whereas in photography the range violet to green is most useful.

It is also customary to express dispersive power more precisely with reference to C, F and D Fraunhofer lines (dark lines) in the solar spectrum. The F, D and C lines lie in the blue, yellow and red regions of the spectrum and their wavelengths are 4861Å and 5893Å and 6563Å respectively. (1Å = 1 Angstrom unit = 10^{-10}m). Thus, the dispersive power may be expressed as

$$\omega = \frac{\mu_F - \mu_C}{\mu_D - 1}$$

The inverse of dispersive power is called as *Abbe's number* or *V-number* and is denoted by V.

$$V = \frac{\mu_D - 1}{\mu_F - \mu_C} \qquad ...(3)$$

A small Abbe's number means high dispersion. Glasses of low dispersion, having V-number above 55, are called *crowns*; and glasses of high dispersion with V-number below 55 are called *flints*.

ACHROMATISM OF SIZE, NOT OF POSITION OF IMAGES

It is to be noted that what is achieved from Eqn. (1) is *not the coincidence of focal points, hut equality of focal lengths for different colours*. To understand this, we note that distance α' of the second principal plane of the combination from the second lens is given for Fraunhofer C and F lines by

$$\alpha_C = -f_C' \frac{t}{f_{1C}'}$$

and $$\alpha_F' = -f_F' \frac{t}{f_{1F}'}$$

Now in achromatism we make $f_C' = f_F'$, but f_{1C}' is not equal to f_{1F}'. Hence the combination *does not have its principal planes H_C' and H_F' in coincidence*. The focal points lie at distance $f_C' = f_F'$ the respective principal planes. Hence the focal points also do not coincide. The principal and focal planes of the image space (those for object space not shown). For a while incident ray the emergent rays of C and F colours are as shown. They *make equal angles with the axis but do not meet the axis at the same point*. An optician does not mind this, because vision depends upon angles subtended. But a photographer will reject the combination.

A hard fact is that no combination of lenses of the same material can achieve achromatism of both positions and magnification of the image. We shall not proceed to give a formal proof of this.

ANGULAR AND CHROMATIC DISPERSIONS

Now we note here the definition of dispersive power as used in *wave theory* of light.

The variation of refractive index μ of a given medium with wavelength is called dispersion. If μ varies rapidly with wavelength, the medium is said to be highly dispersive.

The refractive index μ of the material of a prism is given by

$$\mu = \frac{\sin \frac{A+\theta}{2}}{\sin \frac{A}{2}} \qquad \text{...(1)}$$

where, A is the angle of the prism and θ is the angle of minimum deviation.

$$\sin \frac{A+\theta}{2} = \mu \sin \frac{A}{2}$$

$$\therefore \qquad \cos \frac{A+\theta}{2} = \sqrt{1-\mu^2 \sin^2 \frac{A}{2}}$$

If dθ is the difference in the angle of deviation between two spectral lines of wavelengths λ and λ + dλ, then dθ/dλ is called the *angular dispersion* between the wavelengths λ and λ + dλ.

Differentiating equation (1) with respect to μ on the left-hand side and θ on the right hand side, we obtain

$$d\mu = \frac{\frac{1}{2} \cos \frac{A+\theta}{2}}{\sin \frac{A}{2}} . \, d\theta$$

Dividing by dλ

$$\frac{d\mu}{d\lambda} = \frac{\cos \frac{A+\theta}{2}}{2 \sin \frac{A}{2}} . \frac{d\theta}{d\lambda}$$

$$\therefore \qquad \frac{d\theta}{d\lambda} = \frac{2 \sin \frac{A}{2}}{\cos \frac{A+\theta}{2}} \cdot \frac{d\mu}{d\lambda}$$

$$\text{or} \qquad \frac{d\theta}{d\lambda} = \frac{2 \sin \frac{A}{2}}{\sqrt{1-\mu^2 \sin^2 \frac{A}{2}}} \cdot \frac{d\mu}{d\lambda} \qquad ...(2)$$

$\frac{d\mu}{d\lambda}$ is called *chromatic dispersion* of the material of the prism. It depends on the angle of the prism and refractive index of the material of the prism. In general $\frac{d\mu}{d\lambda}$ is not constant and decreases as λ increases. Using a spectrometer and the given prism, a graph is drawn between μ and λ (μ along the y-axis and λ along the x-axis). The tangent to the curve at any point measures the chromatic dispersion $\frac{d\mu}{d\lambda}$ for that particular wavelength. Substituting this value of $\frac{d\mu}{d\lambda}$ in $\frac{d\theta}{d\lambda}$, can be calculated.

ACHROMATIC COMBINATION OF PRISMS—DEVIATION WITHOUT DISPERSION

An achromatic prism is a combination of two appropriate prisms so constructed that it shows no colours. It is noted that the dispersive powers of different materials are different. Flint glasses have higher dispersive power than crown glasses.

Hence, it is possible to combine two prisms of *different* materials and specified angles such that a ray of white light may pass through the combination *without dispersion*, though it may suffer deviation. Such a combination is called an *achromatic combination.*

An arrangement of two prisms, which does not cause dispersion though it may produce deviation. Let the first prism be made of crown and the second of flint. Let δ_V and δ_R be the deviations and μ_V and μ_R the refractive indices for the violet and red rays in case of crown glass prism. Let the angle of the prism be A. The
angular dispersion produced by the crown glass prism is

$$\delta_V - \delta_R = (\mu_V - \delta_R)A \qquad ...(1)$$

Let, the refracting angle of the second prism of flint glass be A'. It is arranged in such a way that it produces a deviation in a direction opposite to that of the first prism. Let δ_V' δ_R' δ_R' be the deviations and δ_V' and μ_V' the refractive indices for the violet and red rays. The angular dispersion produced by the flint glass prism is

$$\delta_{\varsigma}' - \delta_P' = \left(\mu_V' - \mu_R'\right)A' \qquad ...(2)$$

For achromatism the net angular dispersion produced by the two prisms together must be zero.

$$\therefore \qquad (\mu_V - \mu_R)A + (\mu_V' - \mu_R')\ A' = 0 \qquad ...(3)$$

This is the condition for achromatism.

$$\therefore \qquad A = -\left(\frac{\mu_V' - \mu_R'}{\mu_V - \mu_R}\right)A' \qquad ...(4)$$

The above expression gives the relation between the angles of the two prisms in order that the combination may function as an achromatic combination. The negative sign shows that the two prisms should be placed with their refracting angles in opposite directions. Eqn. (3) may be written as,

$$\frac{(\mu_V - \mu_R)}{\mu - 1}.(\mu - 1)\ A + \frac{(\mu_V' - \mu_R')}{\mu - 1}\ (\mu' - 1)\ A' = 0$$

Using eqn. (6) we can write the above equation as

$$\omega\delta = \omega'\delta' \qquad ...(5)$$

where ω and ω' are the dispersive powers of the two prisms and δ and δ' their mean deviations. Eqn. (5) indicates that the dispersive powers of the prism should be in the inverse ratio of their mean deviation. It is found that when a crown glass prism of refracting angle 60° is combined with a flint glass prism of an angle of 29° 17', the combination becomes achromatic. Though, the combination is achromatic, yet it produces deviation.

PHOTONS

According to quantum theory, light is considerec as a stream of special particles, namely *photons*. Photons do not have rest mass and travel with a velocity equal to the speed of light in a vacuum. The fundamental characteristics of a photon are its energy E and momentum p.

$$E = h\nu \qquad ...(1)$$

and $$p = \frac{h\nu}{c} = \hbar k \quad ...(2)$$

The momentum vector p of a photon has a direction coinciding with that of the wave vector k. k has a magnitude 2 π/λ, and a direction coinciding with that of wave velocity. Photon mass $m = \frac{h\nu}{c^2}$ is the mass of the electromagnetic field and is not associated with a rest mass because photons at rest do not exist.

Fig. 1 : Bursts of light emitted by atoms.

Photons are emitted by individual atoms in light sources. An atom leaving an excited state gives up the excess energy in the form of a burst of light (photon) and jumps to the lower normal state. The process of transition of the atom from an upper state to a lower state lasts for a brief time of about 10^8 8sec. Therefore, the light emitted by an atom is not a continuous harmonic wave of infinite extension but is a *wave train* of finite length having a certain limited number of oscillations. It is impossible to say exactly when an atom may emit light because the emission is completely a random process. Fig. 1 pictorially describes the emission of light by a single atom in terms of wave trains. Other atoms in the source behave similarly but with different emission times. Adding together the wave trains generated by all atoms in the light source produces a succession of wave trains (Fig. 1), which gives an impression of continuous wave.

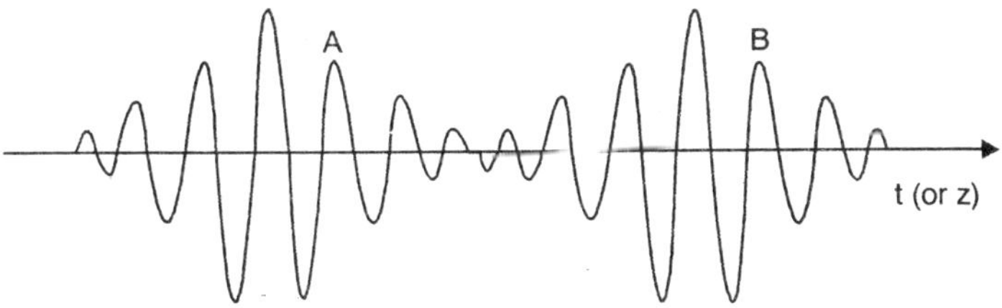

Fig. 2 : A succession of wave trains gives an impression of continuous wave.

DISPERSION WITHOUT DEVIATION

Achromatic prisms are nor of much practical use. A more useful combination of prisms is the one that produces *dispersion but no deviation.* It is called a *direct vision prism.*

It consists of two prisms of different dispersive powers and different angles and combined in such a way that the combination does not produce deviation of the mean rays. If the angles of the crown and flint glass prisms are so adjusted that the deviation produced for the mean rays by the first prism is equal and opposite to that produced by the second prism, then the final beam will be parallel to the incident beam. However, such a combination of two prisms produces dispersion of the incident beam without deviation.

Let us consider two thin prisms, one made of crown glass and the other of flint glass. Let A and A' be the refracting angles of the prisms and μ and μ' the refractive indices. Let δ and δ' be the deviations produced for the mean ray by the two prisms respectively. The deviations are given by

$$\delta = (\delta - 1)A$$

$$\delta' = (\delta' - 1)A'$$

The net deviation produced by the prism combination is to be zero.

That is, $\delta + \delta' = 0$

or $(\mu - 1)A + (\mu' - 1)A' - 0$

$$\therefore \quad A = -\left(\frac{\mu'-1}{\mu-1}\right)A' \qquad \text{...(1)}$$

The negative sign shows that the refracting angles of the two prisms are in opposite direction.

We calculate the net angular dispersion produced by the combination of prisms as follows:

The total deviation for the violet rays is given by

$$\delta_V = (\delta_V - 1)A + (\mu_V' - 1)A'$$

and the total deviation for the red rays is given by

$$\delta_R = (\mu_R - 1)A + (\mu_R' - 1)A'$$

Hence, the total dispersion is equal to

$$\delta_V - \delta_R = (\mu_V - \mu_R)A + (\mu_V' - \mu_R')\,A'$$

Using eqn. (15) into the above equation, we obtain

$$\delta_V - \delta_R = (\mu_V' - \mu_R)A + (\mu_V' - \mu_R')\left(-\frac{\mu-1}{\mu'-1}\right)A$$

$$\text{or} \quad \delta_V - \delta_R = \left[\frac{(\mu_V - \mu_R)}{(\mu-1)} - \frac{(\mu_V' - \mu_R')}{\mu'-1}\right](\mu-1)A$$

It follows from eqn. (6) that

$$\delta_V - \delta_R = (\omega - \omega')\,(\mu - 1)A \qquad ..(2)$$

Thus, the resultant dispersion is equal to the difference between the two dispersions.

DIRECT VISION SPECTROSCOPE

The prisms are fixed in a metal tube. A collimating lens is fixed at one end of the tube and a telescope at the other. The angles of the prisms are such that the total deviation produced for the mean rays is zero. The refracting angles of crown and flint glass prisms are in opposite directions. Thus, the deviation produced by the crown glass prisms in one direction is equal and opposite to that produced by the flint glass prisms in the opposite direction. There will be resultant dispersion and dispersed beam is almost parallel to the incident beam. The prisms are cemented together with Canada balsam ($\mu = 1.54$) to minimize the reflection losses at the interfaces. A spectroscope of this type is very handy and is used to study qualitatively the spectra of different sources of light. Use of more than two prisms increases the resolving power of the instrument, *i.e.*, the spectral lines will appear well separated from one another.

ROTATING MIRROR NULL METHOD

The arrangement used in the later experiments (1926) by Michelson in the famous Mount Wilson measurement of the speed of light. The emphasis is on precision. The light from a slit is collimated by a lens and is reflected by face I of an octagonal mirror on to a concave mirror some 35 km away. A mirror at the focal point of this reflects the beam to the concave mirror and then to face 3 of the octagonal mirror. It is collected by a lens and viewed by the eye, against a fine scale.

When the octagonal mirror is rotated about its axis, face 3 suffers a finite rotation by the time light travels from face 1 to M and to face

3. So the reflected beam shows a shift. But, if the octagonal mirror rotates by exactly one-eighth in this time, face 2 occupies exactly the position of 3 in this time, and hence the spot appears at the same position as when the octagonal mirror is not rotating. The rate of mirror rotation is adjusted for this "null" situation, and measured.

If the "null" comes at n rotation/s, and D is the distance from face 1 to M, we directly get.

$$c = \frac{\text{distance}}{\text{time}} = \frac{2d}{1/8n} = 16\,nD \qquad ...(1)$$

The distance D was measured in two parts. Two marks were provided on two pillars at the two distant stations; their distance (several miles) was measured by the Geodetic Survey of USA. The accepted accuracy of this was 1 part in 200,000. Rest of the distances were measured by Michelson; since these are, after all, small additions to a large distance, even a measurement correct to ± 5 cm was enough.

Correction for refractive index of air was applied by noting the temperature and pressure of air at several points along the 35 km path. The final accepted result for the speed of light in vacuum had a probable error about 1 part in 10,000 largely on account of uncertainty in time measurement and air correction. Later, Michelson used an evacuated tube for the path of light to minimise refractive index correction. Water-pipes were being laid in a town and be utilised a one mile straight section of these pipes with two back and forth reflections and the rotating mirror had 32 faces. The air in the tube was pumped out till pressure was within a few mm. of mercury, and correction for refractive index was made for this residual air. The result was claimed to have a probable error of about 1 part in 30,000. In other words A. A. Michelson, an American Physicist, spent many years of his life in measuring the velocity of light. The method devised by him in the year 1926 at Mount Wilson observatory is considered accurate.

Light from an arc after passing through a narrow slit S is reflected from one face of the octagonal rotating mirror R. Then it is reflected from the small fixed mirrors B and C to a large concave mirror M_1. The point on the mirror C from where light is reflected is the focus of the concave mirror M_1. The light then travels as a parallel beam to another concave mirror M_2, and it is reflected to a plane mirror D at the focus of the concave mirror M_2. The light is then reflected back to the concave mirror M_2, travels back to M_1 and is therefore reflected to a plane mirror F and is then incident on the face A' of the octagonal mirror opposite to A. The final image is viewed through a micrometer eyepiece T with the help of a total reflecting prism P. When the rotating octagonal mirror is stationary, the image is seen by the light reflected from the top surface A'. When it is rotated the image disappears. The speed of rotation of R is suitably adjusted so that the image reappears and is seen in the same position as when R is stationary. The light reflected from A now arrives at A' in the time taken by R to rotate through 45° or 1/8 of a revolution so that the next face is present at A. For this critical speed, the beam is reflected from the next face exactly along the same path as when the mirror is stationary.

If the distance traveled by light in its journey from A to A' is equal to d, then

$$c = \frac{d}{t}$$

If R makes one revolution per second, then

$$t = \frac{1}{8n}$$

$$c = \frac{1}{1/8n} = 8nd \qquad ...(1)$$

Thus, c can be calculated.

Michelson set up M_1 at an observing station on Mt. Wilson and M_2 at Mt. St. Antonio in California, at a distance of 22 miles. The velocity of revolving mirror R was measured by stroboscopic comparison with an electrically maintained tuning fork. He obtained the value of $c = 2.99797 \times 10^8$ m/s.

LABORATORY METHOD FOR FINDING THE VELOCITY OF LIGHT

Kerr cell method is an improvement over the Fizeau's toothed wheel apparatus. It was first devised by Gaviola in 1925. In place of toothed wheel, electro optic shutter is used which is capable of chopping a beam of light several hundred times more rapidly than can be done by the toothed wheel.

ELECTRO OPTIC SHUTTER

It consists of a Kerr cell K placed between two crossed Nicol prisms N_1 and N_2 Kerr cell is a small glass container having two electrodes filled with nitrobenzene. *When a high voltage is applied to the electrodes of K, the light is transmitted through the system. On the other hand, when the field is switched off, light is stopped and not transmitted through the system.*

Thus, by using an electrical oscillator which supplies high frequency voltage, the beam of light can be interrupted at the rate of many millions of times in one second.

Light from a source S after passing through the lens L_1 is made to pass through the nicol prism N_1 and it becomes plane polarized. This beam of light is focused at the centre of the Kerr cell K_1 and falls on the nicol prism N_2. This beam of light is rendered parallel by the lens L_2 and after reflection from the plane mirror M is allowed to fall on the Lens L_3 which concentrates the beam in the middle of the Kerr cell K_2. Finally the beam passes through the nicol prism N_3 and lens L_4.

As the two nicol prisms N_1 and N_2 are crossed, no light reaches the mirror M and the eye. When a high frequency voltage is applied to the Kerr cell K_1 the beam reaches the mirror M and is reflected. As this reflected beam reaches N_2 after passing through K_2, it is again obstructed by N_3 and no light is observed by the eye. It should be remembered that N_2 and N_3 are crossed.

Suppose a high frequency oscillatory voltage is applied to the Kerr cells, K_1 and K_2 simultaneously such that there is no phase difference in their voltages. This means that the two cells will act as two shutters and are allowed to open and close simultaneously. Further, consider the light passing through the cell K_1, when the voltage is maximum and reaching the cell K_2 after some time when the voltage across the cell K_2 is minimum .Then no light reaches the eye.

Thus, we find similar to Fizeau's method, the arrangement N_1, K_1 and N_2 acts as a space for the passage of light while the arrangement of K_2 and N_3 acts as a toothed wheel for the obstruction of the light.

If the distance between the two cells is d and f is the frequency of the H.F. voltage, then time taken by light to travel from K_1 to K_2 is

$$t = \frac{1}{4f}$$

But, $c = \dfrac{d}{t}$

$$\therefore \qquad c = \frac{d}{1/4f} = 4fd \qquad ....(1)$$

In this method, f = 3 × 10^6 Hz and the value of c was found to be 2.99778 × 10^8 ± 20 × 10^3 m/s.

Advantages

(1) As the frequency is very high, this device is capable of chopping a beam of light several hundred times more rapidly than can be done by the toothed wheel. Hence a shorter base line can be used.

(2) The apparatus can be set up in a laboratory.

(3) The accurate frequency of the high frequency oscillator is known.

The Kerr cell KC placed between two nicols N_1, N_2, which are "crossed." The first nicot plane polarises the light from L, and the second does not permit it to pass. The Kerr cell contains two parallel plates immersed in nitrobenzene. If an alternating electrical voltage is applied to the plates, nitrobenzene becomes optically active, in proportion to E field; so it rotates the plane of polarisation of light coming from N_1, and hence N_2 allows a part of this light to pass. In essence we thus have on optical shutter, capable to work at a few megacycles frequency.

Now, the mirrors M_1 to M_6 and the photo-cell R act as follows: The light from L is split by a half-silvered plate M_1. One part goes to mirror M_2. and is reflected back, part of it going to the photoelectric cell R. The other part proceeds on to M_3, and then to the concave mirror M_1 which ren it parallel. The light, then goes to plane mirrors M_6, M_6 and then retraces the path to M_1, part of it reaching R ultimately.

Two beams of light, each with sinusoidal intensity variation produced by the Kerr cell, reach R. The response depends on the time difference in the unequal paths of the two beams. Let T be the period of the sinusoidal variation. If the two beams reach R with time difference 0, T, 2T, etc., the photo-current fluctuates heavily, and hence the amplified current is large. If the two beams reach R with lime difference $\frac{T}{2}, \frac{3T}{2}$, etc., the photo-current does not fluctuate appreciably, and hence the amplified current-is low.

In any experiment we first turn M_4 to an orientation so that it may return the beam on path M_4, M_3, R and adjust M_2 (micrometer motion provided) to get minimum reading in the meter. Then we set M_4 in position and adjust M_2 again to get the minimum reading in the meter.

The additional distance in the seccuad case is M_4 to M_6 and back ± twice the motion of M_2 (other distances are common). Let this additional distance be D. If oscillator frequency is f, we get,

$$c = \frac{Df}{n} \text{ (non integer)} \qquad ...(2)$$

Since the order of magnitude of c is known, there is no difficulty in assigning a proper n, which is generally 1. Anderson used an oscillator of f ~2 × 10^6 cycle per second. Which corresponds to D–150 metres for n = 1. But much higher frequency f can be used, reducing D to laboratory dimension.

ANDERSON'S METHOD

In 1941, W.C. Anderson made use of one Kerr-cell and divided the transmitted light pulses in to two beams by means of a half silvered glass plate. His arrangement is shown in Fig. 11.13. Light from a source S is allowed to pass through a modulated Kerr cell and is split up into two beams by the half silvered glass plate M_1. One beam of the light goes towards M_2 and is reflected back, and the other beam goes along a longer path towards M_3, M_4, M_5 and M_6 and after reflection from M_2 retraces its path and finally the two beams reach the photoelectric cell P.

If the longer path exceeds the shorter path by $\frac{(2n+1)\lambda}{2}$, the resultant of the two modulated waves reaching the photo cell will give a constant intensity. The amplifier connected to the photocell will give zero response in this case.

After this, light after reflection from M_3 is allowed to fall normally on M_4' and it retraces its own path towards M_1. This is done for cutting the light to go to M_5 and M_6. The position of mirror M_2 is adjusted so that again there is zero response in the amplifier connected to the photocell. If the extra path (M_4 to M_6 and back) is a whole number of λ, no change in the response of the photo cell is observed.

In the actual experiment the extra path was found to be equal to 11 λ,.

Here, x = distance from M_1 to M_4 and back,

y = distance from M_1 to M_2 and back,

S = distance from M_4 to M_6 and back,

δx = the distance between M_4 and M_4'

δy = the distance between M_2 and M_4'

In the first case, the path difference is

$$(x + S) - y = (2n + 1)\ \lambda/2 \quad ...(1)$$

and in the second case the path difference is

$$(x + 2\delta\ x) - (y + 2\ \delta y) = \lambda/2 \quad ...(2)$$

Subtracting eqn. (2) from (1), we get

$$S + 2\ \delta y - 2\ \delta = n\lambda \quad ...(3)$$

It was found that n = 11

∴ λ can be known from equation (3).

$$c = \frac{\lambda}{T} \quad ...(4)$$

where, $T = \frac{1}{f}$

Here f is the frequency of H.F. oscillator connected to the Kerr cell and this can be accurately known.

The value of c found by Anderson by this method is 299,776 ± 6 km/s.

This value is the mean of 2895 observations made by Anderson.

Advantages :

(1) The use of two Kerr cells is avoided in the method because it is practically difficult to match the characteristics of the two cells.

(2) The observations are taken by a photo cell instead of the eye.

Source of Error : The main source of error in this method is the difficulty in arranging for both beams to use the same position of the photoelectric cell

PIEZOELECTRIC GRATING METHOD

When a quartz crystal is subjected to a high frequency electric field it is alternately compressed and extended high frequency oscillations are set up. This effect is known as *piezo electric effect*. When these high frequency oscillations are set up in the form of pulses in a quartz crystal, it will be crossed by several parallel nodal planes. In such a case the intensity and refractive index at nodal planes is different from those at other planes and thus the crystal will act as a diffraction grating.

Monochromatic light from a source S is allowed to fall on the quartz crystal C (whose faces are cut parallel) after reflection from the glass plate M. The lens L is used to make the beam parallel. As mentioned, the quartz crystal will act as a grating. Therefore, light after passing through the crystal passes through the slit A of the screen. Finally, it is brought to the focus by the lens L_1. Convex lens L_1 and concave mirror B are fixed at the two ends of a tube such that the length of the tube is equal to the focal length of the lens L_1 and also equal to the radius of curvature of the concave mirror B.

The light retraces its own path and on arriving at the quartz crystal C. If it finds the crystal exactly in the same condition as on its forward journey, then the image of S is seen by the eye. If on the other hand the light on arriving at the crystal C finds that the crystal is not exactly in the same condition as on its forward journey, no image of S is seen by the eye.

Therefore, the quartz crystal C acts similar to a toothed wheel in Fizeau's method or Kerr cell in Anderson's method.

Let the frequency of oscillations be f. Then the rate of formation of the grating is If per second. Therefore, when the image is seen by the eye, the time taken by light to go from C to B and back is a simple multiple of 1/2f.

The tube containing L_1 and B is moved away from the screen along CB. At first it is observed that the image disappears. It again reappears when the distance BB' and back (*i.e.*, 2d) is travelled by light in the time 1/2f second. Several such points can be located when the tube L_1B is moved away from the screen which is d cm apart from each other. For accuracy the mean of all these distances is taken and let it be equal to d.

$$c = \frac{2d}{t}$$

But $$t = \frac{1}{2f}$$

Therefore $$c = \frac{2d}{1} \times \frac{2f}{1} = 4df.$$

In the actual experiment performed by Houston, the value of f was equal to 115×10^6 Hz. The value of c calculated by Houston in vacuum was 299,782 ± 9 km/s, which agrees with the results obtained by other methods.

PROPERTIES OF LIGHT

Reflection, refraction, dispersion and velocity are the important properties of light. We briefly discuss about them here.

Reflection of Light

When light travelling in a medium encounters a boundary leading to a second medium, part of the incident light is returned to the first medium from which it came. This phenomenon is called reflection. Reflection of light from a smooth surface is called regular or *specular reflection.*

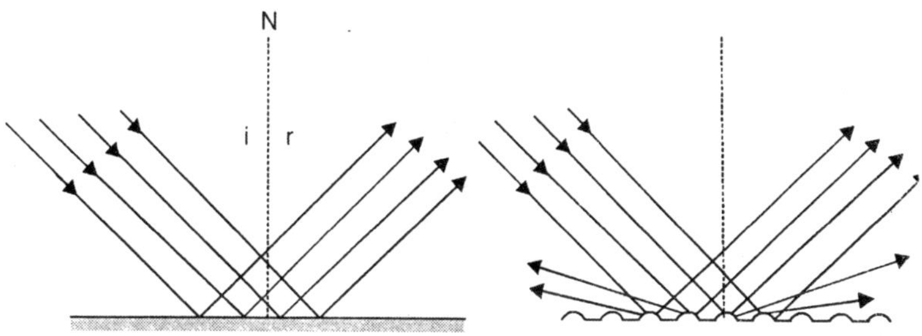

(a) (b)

Fig. 3.

Reflection from a rough surface is known as *diffuse reflection.* It is largely by diffuse reflection that we see non-luminous objects around us. The difference between diffuse and specular reflection is a matter of surface roughness. In the study of optics, the term reflection is used to mean specular reflection.

Laws of Reflection

The light ray AB, passing through air, is incident on a plane mirror and is reflected via the path BC. The point (B) where the light intersects the surface of the mirror is the point of incidence. A line drawn at B, perpendicular to the mirror, is the *surface normal.* The angle subtended by the surface normal and the incident ray is the angle of *incidence*, r. The angle subtended by the surface normal and the reflected ray is the angle of *reflection*, r. For some reason, it is customary to measure the angles from the surface normal toward the ray.

First Law : The incident ray, the reflected ray and the normal at the point of incidence are in the same plane. This plane is called the *plane of incidence.*

Second Law : The angle of reflection is equal to the angle of incidence.

$$i = r \qquad ...(1)$$

Note : The laws of reflection are obeyed in specular reflection. They do not hold in case of irregular or diffuse reflection.

Refraction of LightY

When a ray of light travelling through a transparent medium encounters a boundary leading into another transparent medium, part of the ray is reflected and part of it enters the second medium. The ray that enters the second medium is bent at the boundary and is said to be *refracted.* Thus, refraction means that the light ray follows in the second medium a direction different from its direction in the first medium. The angle 'r' subtended by the normal and the refracted ray is the *angle of refraction.* This angle is also measured from the surface normal toward the ray.

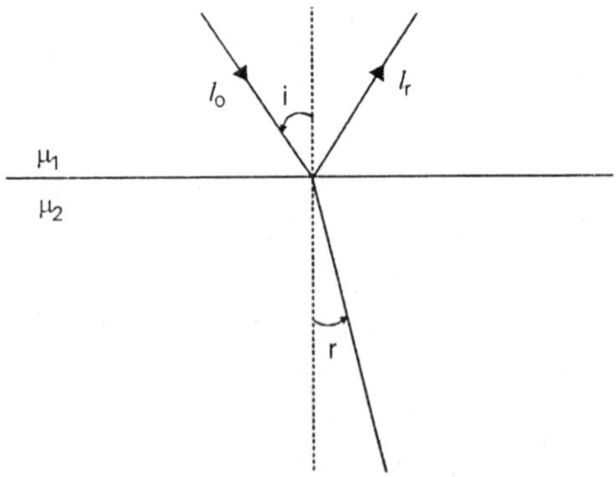

Fig. 4 : Phenomenon of refraction—A ray obliquely incident on air glass interface bends toward the normal in glass.

Laws of Refraction

First Law : The incident ray, the refracted ray and the normal at the point of incidence lie in the same plane.

Second Law : The ratio of the sine of the angle of incidence to the sine of the angle of refraction for any two given media is constant.

$$\frac{\sin i}{\sin r} = \mu \qquad ...(2)$$

where μ is called the refractive index of the medium.

Notes :

(1) The laws of reflection and refraction relate only to the directions of the corresponding rays but do not say anything about the intensities of the reflected and refracted rays. These depend on the angle of incidence, the two indexes of refraction, and the polarization of the incident ray.

(2) The laws of reflection and refraction are obtained as experimental results. They can also be derived from Maxwell's equations. The superiority of this treatment is that it enables us to predict the amplitude, intensity, phase, and polarization states of the reflected and refracted rays.

THE DOPPLER EFFECT

Christian Johann Doppler stated in 1842 that a change of frequency of the wave motion should be observed when a source of sound or light was moving. *The apparent change in the observed frequency of a wave due to the relative motion between the source of waves and the observer is known as the Doppler effect.* The Doppler effect occurs for all types of waves whenever there is relative motion between the source of waves and the observer.

As a train approaches and passes by, there is a distinct change in the sound pitch or frequency of its whistle. The sound gets shriller (higher frequency) as the source approaches a stationary observer and becomes lower in pitch as it recedes. The reason is simple. The source (here the train) radiates spherical waves; but since the source is moving, it emits each successive wave at a different location. As the train approaches the stationary observer, the waves moving in the same direction as the source are *bunched up* in front of the source and towards the observer causing a decrease in wavelength and increase in frequency of the sound heard. As the train moves away from the observer, the sound waves are *spread out* behind the source causing an increase in wavelength and a decrease in frequency of the sound heard. Doppler predicted that the colour of a luminous body must change in the same way by relative motion of the body and the observer.

In the general case where both the source and observer are moving, we obtain the following expression for the *apparent frequency* of the sound heard by the observer.

$$v' = \frac{\upsilon + u_0}{\upsilon - u_s} v \qquad ...(1)$$

where, υ is the velocity of sound in air,

u_0 is the velocity of the observer,

u_s is the velocity of the source of sound, and

v is the actual frequency of the source.

Eqn. (1) is the classical expression for the Doppler shift and includes all possibilities for collinear motion of source and observer relative to the medium. It is seen that the change in frequency due to the Doppler effect is determined by the velocities of the source and the receiver relative to the medium. It is not just the relative motion of source and observer that determines the frequency change. In fact, even when the relative motion is the same, we obtain different quantitative results, depending on whether the source or the observer is moving. This difference occurs because u_0 and u_s are measured relative to the medium in which the sound wave is propagated and because the medium determines the speed.

The Doppler effect also exists for light waves and is important in light. In case of electromagnetic waves there is no medium relative to which the velocity can be defined and hence the Doppler displacement of the frequency of light waves is determined only by the relative velocity of the source and the receiver. This means that 'source receding from observer' and 'observer receding from source' are physically identical situations and must exhibit exactly the same Doppler frequency. The speed of light is so great that only astronomical or atomic sources which have very high velocities show pronounced Doppler effects. There are differences, however, in the Doppler effect formula for light and for sound. The wave speed c is the speed of light and is the same for both source and observer. In the frame of reference in which the observer is at rest, the source is moving away from him with velocity υ. The frequency v' measured by the observer is given by the theory of relativity,

as
$$v' = v\sqrt{\frac{1 - \upsilon/c}{1 + \upsilon/c}} \qquad ...(2)$$

where, υ is the relative velocity of source and observer. When υ is positive, the source moves away from the observer and $v' < v$; when υ is negative, the source moves towards the observer and $v' > v$. This is known as the *Longitudinal Doppler Effect*. It may be noted that the qualitative effect is the same as for sound although the quantitative relationship is different. If a source moves towards the observer the light appears more violet, and if it moves away it appears more red.

There is a further consequence of relativity on the Doppler effect. Whereas in the classical case there is no frequency shift if the relative velocity of source and receiver is perpendicular to the line joining them (for example, when the source travels along a circle at whose centre is located the receiver), that is no longer true in the relativistic case. In addition to the longitudinal effect, a *Transverse Doppler Effect* exists for light waves. The transverse Doppler effect is a strictly relativistic phenomenon and is a consequence of time dilation. It consists in a reduction in the frequency picked up by the receiver observed when the vector of the relative velocity is directed at right angles to the straight line passing through the receiver and the source. In this case, the frequency v in the frame of the source is associated with the frequency v' in the frame of the receiver by the relation

$$v' = v\sqrt{1-u^2/c^2} \approx v\left[1-\frac{1}{2}\cdot\frac{\upsilon^2}{c^2}\right] \qquad \text{...(3)}$$

The relative change in frequency in the Transverse Doppler Effect, given by,

$$\frac{\Delta v}{v} = -\frac{1}{2}\frac{\upsilon^2}{c^2} \qquad \text{...(4)}$$

is proportional to the square of the ratio υ/c and is consequently considerably smaller than in the longitudinal effect for which the relative change in the frequency is proportional to the first power of υ/c. The existence of the transverse Doppler effect was proved experimentally by the American physicists H.E. Ives and G.R. Stilwell in 1938.

The longitudinal Doppler effect is used to determine the speeds at which luminous heavenly bodies are moving towards us or receding from us. Analysis of the spectra of light from distant stars shows shifts in wavelength compared to spectra of the same elements from the stationary light sources here on the earth. These can be interpreted as Doppler shifts due to motion of the stars. The shift is nearly always toward the longer wavelength or red end of the spectrum and is therefore called the *red shift*. Such Doppler red shift shows that the galaxies are moving away from us, the recession velocity being greater for the more distant galaxies. These observations are the basis of the concept of "expanding universe". According to *"Big Bang"* theory, all the matter in the Universe was in one big mass and a big bang occurred in the remote past. Since then the bodies have been moving away. Since we can calculate the expansion rate of the galaxies, it is estimated that the expansion of the Universe began about 15 billion years ago.

The thermal motion of the molecules of a luminous gas, owing to the Doppler effect, leads to *broadening of the spectral lines*. As a result of the chaotic nature of the thermal motion, all the directions of the molecular velocities relative to a spectrograph are equally probable. Therefore, the radiation registered by the instrument contains all the frequencies in the interval from v_0 $(1 - \upsilon/c)$ to v_0 $(1 + \upsilon/c)$, where v_0 is the frequency emitted by the molecules, and υ is the velocity of thermal motion. The width of a recorded spectral line is thus $2v_0\ \upsilon/c$. The magnitude of the Doppler broadening of spectral lines makes it possible to assess the velocity of thermal motion of the molecules and consequently, the temperature of a luminous gas.

The Doppler effect provides a convenient means of tracking a satellite, of measuring the speed of an aeroplane, or the speed of an auto-mobile. The satellite emits a radio signal of constant frequency v. The frequency of the signal received on earth decreases as the satellite is passing over. If the received signal is combined with a constant signal generated in the receiver, it produces beats. The beat frequency produces an audible note whose pitch decreases as the satellite passes overhead. To measure the speed of speeding cars, a generator of electromagnetic waves is located usually in a patrol car stationed at the side of the road. The wave is reflected from a moving car, which thus acts as a moving source. The reflected wave is Doppler-shifted in frequency. Measurement of the frequency shift using beats permits the measurement of the speed of the car.

DEFINITION OF LIGHT

Light is the form of invisible energy which produces in us the sensation of sight. We can see the object in the presence of light at through light itself is invisible. We understand the world around us with the help of information reaching our five sense instruments namely eyes, ears, tongue, nose and skin as vision (sight). Nothing has been so elusive or so secretive than light. Man could gain an understanding about the nature of light after many centuries of persistent efforts. Light is the agent which stimulated our sense of sight. The eye convert the incoming light into electrical signals and convey them to the brain. Which after processing the signals causes images or picture to be created in our mind.

We study the light is one of the most important part of Physics since indeed around us is gained through seeing. We learnt about the properties of giant stellar systems by means of light that travels for millions of years through empty space to deliver us its massage.

We briefly answer these questions in this introductory chapter and elaborate them at appropriate places in the book. Let us start with a look at some of the important landmarks in the evolution of our understanding about light.

Determination of velocity of light is of great importance. Einstein has shown that the energy released by the nucleus of an atom or otherwise is given by $E = mc^2$; where m is the decrease in mass and c is the velocity of light.

But, before the 17th century, it was thought that velocity of light is infinite. The fact that the flash of lightening is seen instantaneously and the sound is heard after some time shows that velocity of light is greater than the velocity of sound.

The first attempt to find the velocity of light was made by Galileo in 1600.

GALILEO'S EXPERIMENT

Galileo suggested the following method for finding the speed of light: Two men with lanterns stood over two hills, a measured distance apart. The first uncovered his lantern and started a clock, and the second uncovered his lantern on seeing the light from the first. When the first man saw the second man's light, he stopped the clock. The experiment failed. But, it was not a complete failure because it showed that the speed of light was too great for measurement with short distances and crude timings.

These days laser *microwave pulses* are reflected from the moon (distance 8×10^8 m), and from the time of return of the pulse one may get the speed of light c. But other methods have given c so accurately that these reflections are now used to measure moon's distance and its variations.

Roemer in 1675 used a method not needing small time measurement The satellites of planet Jupiter show eclipses once in every 7 days which is their time of revolution about Jupiter. But the period was found by Roemer to be 11 minutes longer than the mean when earth was farthest from Jupiter and 11 minutes shorter when earth was nearest. Roemer attributed this 22 minutes difference to the time taken by light in travelling the diameter of earth's orbit. This diameter being 3×10^{11}m, Roemer calculated the speed c:

$$c = \frac{3 \times 10^{11}\,\text{m}}{22 \times 60\,\text{sec}} = 2.3 \times 10^{8}\ \text{m/sec}.$$

Later measurement showed that the time difference was actually 16 min. 20 sec., and hence the value needed revision. But that is not the point A great step had been taken in establishing that speed of light, though large, is *not infinite.*

Haifa century later, in 1726, *Bradley* noted that the fixed stars showed apparent displacements (aberrations) in angular positions over the year, and he attributed it to the finite speed of light, obtaining a value in agreement with that obtained by Roomer's theory. Both these may be listed as astronomical methods.

In fact the word 'method' is probably not appropriate. A natural observation was explained by an assumption in each case. The assumption may or may not be valid. For instance, one could at least imagine that the moons of Jupiter really did not go round Jupiter with constant period over earth's year for some reason. To say that the periods are not likely to vary and the differences in our observations are due to the extra distance 3×10^{11} in travelled by light when we are farthest from Jupiter, is a theory. The best support for it was that Bradley, in interpreting another astronomical phenomenon, got a value of c, in close agreement with Roemer's.

The first *direct* measurement of the speed of light from an experiment of sending a signal over a distance and measuring the time of travel came over a century after Bradley's work. That was in 1849. The experimenter was *Fizeau* using path lengths ~ 10 miles. Soon after that *Foucautt* discovered a method which involved distances of only a few metres and which was therefore suitable for measuring speeds of light in material media. Another *terrestrial method* was devised by Michelson, around 1880-83 using long paths to get a high precision determination of c. An entirely different method was later discovered which used path lengths of only a few metres and yet gave accuracy of the same order as Michelson's. It is the Kerr cell method, devised in 1925 by *Karolus* and *Mittelstaedt* and improved, among others, by *Anderson.*

Two observers were stationed at a distance of a few-kilometers. One observer uncovered his lamp and the second observer uncovered his lamp after seeing light from the lamp of the first observer. The first observer tried to measure the time interval between the uncovering of his lamp and the light seen from the lamp of the second observer. If the distance between the two observers = x, then $c = \frac{2x}{t}$. But Galileo failed to find the velocity of light as the time interval t was very small and could not be measured accurately.

The first successful attempt was made by Romer, a Danish astronomer in 1676.

ROMER'S ASTRONOMICAL METHOD

Romer observed the eclipses of the Jupiter's satellites at times when earth was at different positions with respect to Jupiter. He found that while the earth, in its orbital motion round the sun, receding from Jupiter, the mean period between two successive eclipses of a particular satellite is longer than that when the earth is moving nearer the Jupiter. This anomally formed the basis for the calculation of velocity of light. He explained when the earth is receding from Jupiter; light has to travel a greater distance at each successive disappearance of satellite whereas when the earth is approaching the Jupiter, light has to travel a shorter distance at each successive disappearance of the satellite.

Jupiter has a number of satellites or moons revolving round it. Jupiter makes a complete revolution around the sun in 11.86 years whereas the earth completes one revolution in one year. It is assumed for the sake of simplicity that the orbit of the earth and Jupiter are circular.

The satellites which revolve round the Jupiter have their periods lying between 11 hours 58 minutes for the satellite nearest the planet and 16 days, 16 hours, 32 minutes and 11 seconds for the most remote satellite.

As the satellites revolve in orbits nearly parallel to the plane of the Jupiter's orbit, each satellite, once in every revolution, enters the shadow of the Jupiter and so becomes eclipsed. Romer studied the eclipse of the innermost satellite of Jupiter. At some time, Jupiter J, and the earth E_1 are on the same side of the sun and are in conjunction. If light were transmitted instantaneously, the actual time of eclipse and its observation on the earth should be the same. If light has a velocity c, then light from the satellite at the time of eclipse has to travel a distance $J_1 E_1$ before reaching the earth. Thus, the eclipse will be observed at a time J_1E_1/c later on the earth. If the actual time of the eclipse is T_2 and T_1 is the time when it is observed on the earth then

$$T_1 = T_0 + J_1R_1/c \qquad ...(1)$$

After a lapse of 0.545 of a year, when the earth is at E_2 the Jupiter will be at J_2 *i.e.* in opposition. Let the eclipse of the same satellite be observed at this time. It will be observed J_2E_2/c seconds after its actual occurrence. If t is the time for one revolution of the satellite the time actually elapsed between the first and the nth eclipse will be = (n – 1)t. If T_2 is the time of observation of the eclipse on the earth, then,

$$T_2 = T_0 + (n - 1)t + \frac{J_2E_2}{c}. \qquad ...(2)$$

Subtracting (1) from (2), we get

$$T_2 - T_1 = (n - 1)f + \frac{J_2E_2 - J_1E_1}{c} \qquad ...(3)$$

$$\text{Let } T = T_2 - T_1 = (n - 1)\,t + \frac{J_2E_2 - J_1E_1}{c}$$

$$= (n-1)\,t + \frac{d}{c}$$

where d is the diameter of the earth's orbit round the sun

$$T = (n - 1)\,t + \frac{d}{c} \qquad ...(4)$$

Similarly, it can be proved that the time interval between the first and the nth eclipse observed on the earth when the earth moves from E_2 to E_3 will be

$$T' = (n - 1)t - \frac{d}{c} \qquad ...(5)$$

i.e., when the earth moves from opposition to conjunction.

Subtracting (5) from (4), we obtain

$$T - T' = \frac{2d}{c}$$

Romer observed that the time interval T was 32 minutes 52 seconds more than the time interval T'. Taking the diameter of the earth's orbit as 185.6 × 106 miles and the time interval T – T' = 1972 seconds, Romer calculated the value of c as

$$c = \frac{2 \times 185.6 \times 10^6}{1972} \text{ miles/sec ond}$$

$$c = 186{,}000 \text{ miles/second (approx.)}$$

This method is not very accurate due to the following reasons:

(i) orbits are not circular but they are elliptical

(ii) correct value of the diameter of the earth is not known

(iii) it may not be exact time when the eclipse occurs.

ABERRATION METHOD BY BRADLEY'S

This method is based on the principle of relative velocity. When rain is falling vertically downward, a person standing on the road holds his umbrella vertically, but when he moves on the road, he holds his umbrella in a direction making some angle with the vertical.

This fact has been used in the measurement of the velocity of light by Bradley.

In 1729, Bradley observed that the angular elevation of a 'fixed' star varied slightly according to the position of the earth in its orbit around the sun. One day he was being rowed across a stream in a boat and noticed that the boat drifted slightly downstream. He at once thought that the difference between the actual and the observed angular elevation of the star was due to a combination of the velocity of the earth in its orbit and that of the velocity of light.

If the earth were stationary, the telescope would have to point in the direction AS analogous to umbrella when the man is stationary to observe the star. But since the earth is moving with a velocity of 18.5 miles/second around the sun, the telescope would have to be directed along AS' to observe the star. The direction of AS' is that of the relative velocity of light with respect to the earth. The angle between the true and me apparent direction of the star is α.

Here, $\tan \alpha = \upsilon/c$

Since the angle is small, $\alpha = \upsilon/c$

(i) When the earth is stationary;

(ii) When the earth is moving in its orbit;

(iii) Relative velocity of light with respect to the velocity of the earth.

The earth is at E_1 it is moving towards the star, and at E_3 it is moving away from the star. In these cases for the purpose of receiving light from the star, it is stationary and there is no aberration. The aberration is maximum at E_2 or E_4 because the earth will be moving normal to true direction of the star (analogous to the man moving normal to the direction of the rain). The direction of the motion of the earth is continually changing and the earth is moving in diametrically opposite directions at the spring and autumn equinoxes. Therefore, on observing the position of the star near the zenith at these times, through a telescope it will be seen to occupy different positions with respect to the horizon. At the spring equinox, $\alpha = \alpha/c$ and at the autumn equinox, $\alpha = \upsilon/c$.

From the two positions of the telescope, the aberration can be calculated and it is equal to half the angle between the two positions of the telescope.

The angle α was found to be 20.24" of an arc and since $\upsilon = 28.5$ miles/second the value of the velocity of light c was calculated as 186,230 miles/second. However, aberration is independent of the distance of the star from the earth and should not be confused with parallax.

FIZEAU'S METHOD

The first terrestrial method for determining the velocity of light was performed by Fizeau in 1849. The experimental arrangement A bright source S emits light which after passing through a lens L and after reflection from the plane P is converged to a point H. The point H lies at the focus of the lens L_1 and in a space between the two teeth of the wheel W. Therefore, the light after passing through L, is rendered parallel and after traveling a distance of few miles is allowed to fall on the lens L_2.

The light after refraction through the lens L_2 is brought to focus at A which is also the pole of the concave mirror M. The radius of curvature of the mirror M is equal to twice the focal length of the lens L_2. Due to this the rays are reflected along their original path. An image of S can thus be observed by the eye placed behind the eyepiece E.

The rim of the toothed wheel W is at H and the wheel is rotated about a horizontal axis. This is the important path of Fizeau's experiment. The teeth and spaces of the wheel are of equal width. Fizeau used a wheel with 720 teeth.

Working : When the wheel W is rotated, an image of the source is observed through E as long as light passes through the space of the wheel towards E. The wheel is rotated at such a speed that the rays passing the space fall on the tooth on their return. When this is adjusted, then no image of the source S is seen by the eye. If the speed of the rotation of the wheel is doubled, again the image of the source S is seen by the eye because the rays on their return journey again pass through the next space. When the speed of the wheel is three times the first, no image of the source is seen by the eye because the rays on the return fall on the next tooth. In this way, if the speed of the wheel is continuously increased, the image of the source S will be alternately blocked and seen.

If the distance between H and A is equal to d and the wheel makes n rotations per second when the first eclipse is observed, then the time taken by light to travel from H to A and back is

$$t = \frac{2d}{c}$$

If the wheel has 'm' teeth, m spaces and in the time t, the wheel moves from the centre of the space to the centre of the tooth, then

$$t = \frac{1}{2} m n$$

$$\frac{2d}{c} = \frac{1}{2} m n$$

and $$c = 4 m n d \qquad \text{...(1)}$$

If n_2 is the number of rotations per second of the wheel when the second eclipse is observed

then $c = 4$ m nd/3, because $n_2 = 3$ n. In Fizeau's actual experiment

$$n = 12.6 \text{ rotations/second}$$

$$m = 720$$

$$d = 8633 \text{ meters}$$

$$c = 4 m n d$$

$$c = 3.13 \times 10^8 \text{ m/s}$$

The main advantage of this method is that the principle involved is simple and Fizeau actually took the idea from the experiment attempted by Galileo (covering and uncovering the lamp).

It should be remembered that Fizeau's experiment is not free from criticism due to the following reasons:

(1) The complete eclipse or disappearance of light cannot be obtained due to scattering of light from the teeth.

(2) The image of the source is very faint because the intensity of light is considerably decreased due to the refraction and reflection at various surfaces of the lenses and mirror.

(3) Uniform speed of rotation of the wheel cannot be attained.

(4) The appearance or disappearance is not abrupt but it takes place gradually from maximum to minimum and *vice versa.*

Improvement:

(1) Comu in 1874 determined the velocity of light by Fizeau's method with improved apparatus. He used a distance of 23 kilometers and instead of determining the velocity of rotation of the wheel for the disappearance of the image, he determined the velocity for which the brightness of the image, appears to become minimum and where the image begins to increase in brightness. He took the mean of these velocities as the true velocity of rotation of the wheel for the purpose of calculation. His result for the velocity of light in air is 3.004×10^8 m/s.

(2) Young and Forbes in 1881 determined the velocity of light by Fizeau's method and beveled the teeth of the wheel so that light stopped by the teeth of the wheel was reflected to the sides. Moreover, they used the silvered plates having a small aperture for viewing the image instead of the glass plate P. They calculated the value of c as 3.013×10^8 m/s.

(3) Perrotin used Fizeau's method and kept the distance equal to 40 kilometers and calculated the value of the velocity of light in air as 2.999×10^8 m/s.

In other words fact the heart of the method is the toothed wheel. The teeth and the gaps in between made an alternating shutter when the wheel was made to rotate. In one of the experiments the wheel had –500 teeth and rotated at –10 rev/sec. Thus time intervals of 1/5000 sec were obtained. Now, the image of a strong source A is formed at X by the lens L_1 and 45° inclined half-silvered glass plate P. X is at the rim of the toothed wheel W, described above. If there is a gap at X, light passes through, is rendered parallel by lens L_2, and after travelling distance of several kilometres falls on the system L_3M. The focal point of lens L_3 lies on mirror M and the centre of curvature of M lies on L_3. The net result is that the beam is returned by the system on its own path and is collected by lens L_2 at X. If there is a gap in wheel W *at that moment*, the light passes through. A part of this light it passed by the plate P, and is collected by lens L_4 at B. This image is sees through the eyepiece E.

All the four lenses are compound achromatic systems; they are shown in the figure as single lenses for clarity only.

To start an experiment, a 'gap' in W is brought at X and optical setting is completed to get the return image of A at B. Then the wheel is rotated and its speed is gradually increased. If the time taken by light to travel distance AM = d and back is equal to the time in which the wheel rotates to replace a 'gap' by a 'tooth' at X, then the image at it is cut off or 'eclipsed.' At this stage, the rotation speed of wheel is kept constant and the number of revolutions per second (m) is measured.

If the number of teeth in W is n, we have

$$c = \frac{\text{distance}}{\text{time}} = \frac{2d}{\dfrac{1}{(2mn)}} = 4\text{ mnd} \qquad \text{...(2)}$$

If the rotation of IV is made faster still, a stage comes where the image 5 becomes bright again. It can be at once seen that if this time the number of revolutions of W per second is m_1, we have

$$c = 2m_1nd; \qquad \text{...(3)}$$

and so on.

It is to be realised that eclipse or full brightness condition do not occur suddenly; as the speed is gradually increased, the intensity goes on calling till a total cclipse occurs.

THE DUAL NATURE

On one hand light appears to be a continuous electromagnetic wave of frequency v and on the other hand it appears to be a collection of photons having energy E and momentum p. It has been found that neither of the models can separately explain *all* the experimental facts. A particle is precisely localized in space whereas a continuous wave cannot be attributed to a particular location in space. Thus, the corpuscular nature and wave nature appear to be mutually exclusive. However, the experimental evidence points out that light behaves both as a continuous wave and a particle. Therefore, we say that light exhibits *dual nature.* The reason for the wave-particle dualism may be understood as follows. At the lower frequency end of the electromagnetic spectrum are radio waves whose wavelengths are so large that they spread over a very large volume of space (Fig. 5).

Therefore, the energy available at any point is insignificantly small and their particle nature cannot be observed. On the higher frequency side of the spectrum, we have x-rays and γ-rays. Their wavelengths are so short that the wave energy is concentrated in a point of very small dimension and the particle nature is readily experienced whereas the wave nature is less noticeable. The visible region represents the transition region in which both the aspects of light can be observed.

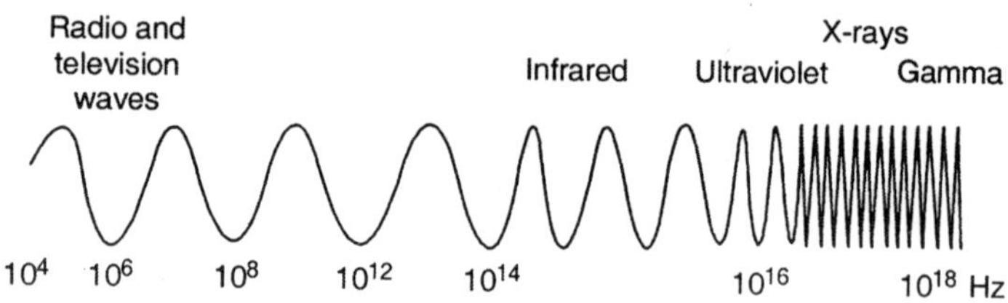

Fig. 5.

In photon description the light intensity is given by I = N hv and in the wave description it is given by $7 = |E|^2$. It implies that the square of the amplitude of a light wave at a point in space is proportional to the number of photons arriving at that point. In other words the amplitude of a light wave determines the probability that a photon can be found at a particular point in space. Probability to observe photons is thus proportional to $|E|^2$. Thus, the ultimate connection between the wave behaviour and the particle behaviour are interpreted.

To sum up, light *rays* consist of streams of *photons*, which are basically *electromagnetic wave trains.* We use all the three descriptions namely rays, waves and photons appropriately to describe the behaviour of light. When light is refracted or reflected by optical elements such as lenses, we mostly use the ray description; when light propagates through space or any medium we use the electromagnetic wave description and whenever light interacts with matter we make use of the photon description.

HISTORY OF THE NATURE OF LIGHT

Around 1836, Faraday (1791-1867) showed that a varying magnetic field induces an electromotive force and thus established the intimate connection between electricity and magnetism. Further, Faraday showed that the polarization of light was affected by a strong magnetic field, which was the first hint as to the electromagnetic nature of light. Clerk Maxwell (1831-1879) unified the empirical laws of electricity and magnetism into a coherent theory of electromagnetism. In 1873, Maxwell showed that the speed of electromagnetic waves equals the speed of light. On the strength of this, he made the prediction that light is a high frequency electromagnetic wave. In 1887, Hertz (1857-1894) confirmed Maxwell's theoretical prediction by producing and detecting electromagnetic waves. The electromagnetic waves were initially supposed to be supported by the ether medium. Though, electromagnetic theory is capable of explaining the phenomena connected with the propagation of light, it fails to explain the processes of emission and absorption. H.A. Lorentz (1853-1928) assumed that ether is in a state of absolute rest to be the carrier of electromagnetic field.

In 1887, Michelson-Morley performed the famous ether-drift experiment and found that light travels at the same speed irrespective of the position of the earth in its orbit. It led to the conclusion that ether does not exist. Hence, *light is a self-sustaining high frequency electromagnetic wave.* This theory is known as the *Field Theory.*

DEVELOPMENT OF GEOMETRIC OPTICS

The Greeks were aware of the rectilinear propagation of light. They knew that when light is reflected from a mirror, the angle of incidence is equal to the angle of reflection. This was stated by Euclid (300 B.C.) in his book Catoptrics. Hero of Alexandria suggested that light traverses the shortest path between two points. They were also aware of refraction of light as it passes from one transparent medium to another. Claudius Ptolemy (130 A.D.) of Alexandria measured the angles of incidence and refraction for several media. Further progress came to a halt with the fall of Roman Empire in 475 A.D. Study of light was again revived in Europe during the thirteenth century. Francis Bacon (1215-1294) suggested the idea of using lenses to improve eyesight. In about 1280, spectacle lenses came into use to correct faulty vision. In 1609 Galileo (1564-1642) devised a practical telescope. Van Leeuwenhoek (1632-1723) developed the first microscope. John Kepler discovered the phenomenon of total internal reflection. In 1621 Willebrod Snell (1591- 1626) and independently in 1637 Rene Descartes (1596-1650) discovered the law of refraction. In 1658, Fermat (1601-1655) discovered the principle of least time. According to this principle, light always follows that path which takes it to its destination in the shortest time. Here derived the law of reflection and refraction applying this principle of least time for the path followed by light. In 1660 the phenomenon of diffraction was noticed by Grimaldi (1618-1663). In 1667 Newton established that white light is composed of seven independent colours. In 1670, Bartholinus (1625-1698) discovered the phenomenon of double refraction.

In 1675 Isaac Newton (1642-1727) put forward the corpuscular theory. According to this theory, a luminous body emits in all directions streams of extremely minute particles, called *corpuscles.* They are supposed to travel through a medium with a tremendous but finite velocity in straight line paths.

The particle theory of Newton could explain the straight line propagation of light and that an object casts a sharp shadow; but it failed to explain why the continued loss of particles did not also cause a source of light to lose weight. However, the theory could prove the laws of reflection and refraction of light. Newton predicted that light should travel faster in a denser medium than in a rarer medium. However, the phenomenon of diffraction and Newton's rings could not be explained on the basis of corpuscular theory.

In 1676, Romer (1644-1710) proved that light travels with a finite velocity. Robert Hooke (1635-1703) studied the coloured patterns formed due to thin film interference.

DEVELOPMENT OF WAVE OPTICS

In 1678 Huygens (1629-1695), a contemporary of Newton, proposed wave theory of light. According to this theory, light energy is supposed to be transferred from one point to another in the form of waves. Huygens was able to prove the ordinary laws of reflection and refraction. He predicted that light should travel slower in a denser medium than in a rarer medium. He also explained the phenomenon of double refraction by assuming two types of waves. The wave theory was not accepted immediately. The chief reason was that a wave motion needs a medium; but light could travel to us from the sun through the vacuum of space.

In 1803, Thomas Young (1773-1829) demonstrated for the first time the interference of light beams. He also explained Newton's rings and the colours of thin films on the basis of interference of light waves. Thomas Young provided strong support to the wave theory. In 1808, Malus (1775-1812) discovered the polarization of light. In 1815, Augustin Fresnel (1788-1827) further developed the wave theory and explained the rectilinear propagation of light which has been the chief Obstacle in the way of accepting wave theory. He provided a satisfactory explanation of the diffraction phenomenon.

Following Huygens, both Young and Fresnel assumed that light waves are longitudinal. Young and Fresnel conceived of an elastic medium, which was assumed to exist pervading the entire universe, and it was named luminiferous ether. The vibrations of the ether propagated as light, just as longitudinal vibrations in air propagate as sound. But the longitudinal wave theory of light could not explain polarization, a property exhibited by transverse waves but not by longitudinal waves. Fresnel and Arago (1786-1853) conducted experiments on super-position of linearly polarized light. Young eventually realized that light is a transverse wave and in 1817 explained the results of Fresnel and Arago's experiments. In 1850, Jean Foucault (1791-1868) established that light travels slower in liquids than in air. This is just opposite to the prediction of Newton's theory.

Finally, the wave model was accepted. The acceptance of the wave theory of light made it obvious that a supporting medium should exist. Subsequently, elastic ether theory was developed during the next ten years. Strange properties were attributed to it. It was assumed to be extremely rigid so that it can support the exceedingly high frequency oscillations of light travelling at a speed of 3×10^8 m/s; yet it does not offer resistance to the motion of celestial bodies through it. Its density was supposed to increase in material substances to account for the lower velocity. In 1823, Presnel derived expressions for the reflection and transmission coefficients on the basis of ether theory.

DEVELOPMENT OF QUANTUM OPTICS

In 1814 Fraunhofer discovered dark lines in the solar spectrum. In 1861 Bunsen and Kirchhoff attributed them to the absorption of certain wavelengths by the gases in the outer atmosphere of the sun. It was also found that every gaseous chemical element possesses a characteristic line spectrum. The detailed studies of emission and absorption spectra of elements evolved into a separate discipline. In 1900, in order to obtain a correct theoretical expression for the black body radiation. Max Planck (1858-1947) found it necessary to suppose that light is absorbed or emitted in the form of elementary quanta. In 1905, Einstein (1879-1955) made use of the quantum concept to successfully explain the photoelectric emission. According to him, light is a stream of photons. In 1913, applying Planck's quantum hypothesis, Niels Bohr (1885-1962) devised an atomic model for the emission and absorption of light. It successfully explained the simple laws of line spectra of gases. The traditional sources of light produce incoherent light. The first coherent source of light, namely laser was built in 1960. The high power lasers led to a number of nonlinear optical effects such as harmonic generation, frequency mixing etc. Quick developments in holography and fibre optics followed the discovery of lasers.

We now visualize a photon as a bundle of electromagnetic radiation that oscillates with a definite frequency and travels through free space with the speed of light. Individual photons carry energy and momentum, so light has particle-like properties. When the number of identical photons is very large, they exhibit the properties of a continuous wave with the same definite frequency and propagation speed as the quantum. The phenomena of interference, diffraction and polarization and propagation of light in space is adequately explained by classical electromagnetic wave theory, whereas the experiments involving interaction of light with matter, such as photoelectric effect are best explained by assuming that light is a particle.

THE FOUR IMPORTANT THEORIES

Various theories have been put forward about the nature of light. We will make a brief survey of the four important theories which guided the evolution of our understanding of the nature of light. The theories are known as :

1. Corpuscular theory
2. Wave theory
3. Electromagnetic theory and

4. Quantum theory.

CORPUSCULAR THEORY

The corpuscular theory was postulated by ancient Greeks and was favoured by Sir Isaac Newton. According to this theory, a luminous body continuously emits tiny, light and elastic particles called *Corpuscles* in all directions. These particles or corpuscles are so small that they can readily travel through the interstices of the particles of matter with the velocity of light and they possess the property of reflection from a polished surface or transmission through a transparent medium. When these particles fall on the retina of the eye, they produce the sensation of vision. On the basis of this theory, phenomena like rectilinear propagation, reflection and refraction could be accounted for, satisfactorily. Since the particles are emitted with high speed from a luminous body, they, in the absence of other forces, travel in straight lines according to Newton's second law of motion. This explains rectilinear propagation of light.

Reflection of Light on Corpuscular Theory

Let SS' be a reflecting surface and IM the path of a light corpuscle approaching the surface SS'. When the corpuscle comes within a very small distance from the surface (indicated by the dotted line AB) it, according to the theory begins to experience a force of repulsion due to the surface (Fig. 6).

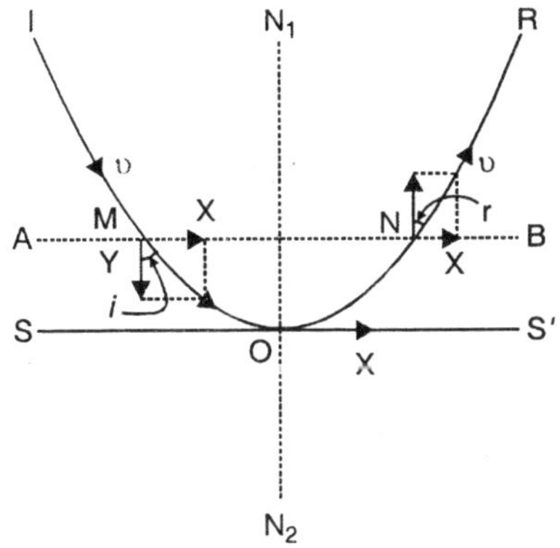

Fig. 6

The velocity υ of the corpuscle at M can be resolved into two components x and y parallel and perpendicular to the reflecting surface. The force of repulsion acts perpendicular to the surface SS' and consequently the component y decreases up to O and becomes zero at O the point of incidence on the surface SS'. Beyond O, the perpendicular component of the velocity increases up to N. Its magnitude will be again y at N but in the opposite direction. The parallel component x remains the same throughout. Thus, at N, the corpuscle again possesses two components of velocity x and y and the resultant direction of the corpuscle is along NR. The velocity of the corpuscle will be υ. Between the surfaces AB and SS' the path of the corpuscle is convex to the reflecting surface. Beyond the point N, the particle moves unaffected by the presence of the surface SS'.

$$x = \upsilon \sin i = \upsilon \sin r, \qquad (\because i = r)$$

Further, the angles between the incident and the reflected paths of the corpuscles with the normals at M and N are equal. Also, the incident and the reflected path of the corpuscle and the normal lie in the same plane viz., the plane of the paper.

Refraction of Light on Corpuscular Theory

Newton assumed that when a light corpuscle comes within a very small limiting distance from the refracting surface, it begins to experience a force of attraction towards the surface.

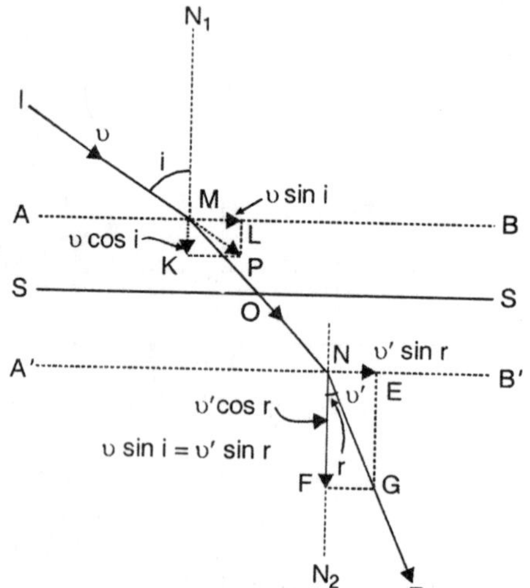

Fig. 7.

Consequently the component of the velocity perpendicular to the surface increases gradually from AB to A'B'. SS' is the surface separating the two media (Fig. 7) I M is the incident path of the corpuscle travelling in the first medium with a velocity υ and incident at an angle i. AB to A' B' is a narrow region within which the corpuscle experiences a force of attraction. NR is the refracted path of the corpuscle. Let υ sin i and υ cos i be the components of the velocity of the corpuscle at M parallel and perpendicular to the surface. The velocity parallel to the surface increases by an amount which is independent of the angle of incidence, but which is different for different materials. Let υ and υ' be the velocity of the corpuscle in the two media and r the angle of refraction in the second medium.

As the parallel component of the velocity remains the same,

$$\upsilon \sin i = \upsilon' \sin r$$

or

$$\frac{\sin i}{\sin r} = \frac{\upsilon'}{\upsilon}$$

$$= \frac{\text{velocity of light in the second medium}}{\text{velocity of light in the first medium}} = {}_1\mu_2$$

= (refractive index of the second medium with reference to the first medium)

Thus, the sine of the angle of incidence bears a constant ratio to the sine of the angle of refraction. This is the well known Snell's law of refraction. If $i > r$, then $\upsilon' > \upsilon$ *i.e.*, the velocity of light in a denser medium like water or glass is greater than that in a rarer medium such as air.

But the results of Foucault and Michelson on the velocity of light showed that the velocity of light in a denser medium is less than that in a rarer medium. Newton's corpuscular theory is thus untenable. This is not the only ground on which Newton's theory is invalid. In the year 1800, Young discovered the phenomenon of interference of light. He experimentally demonstrated that under certain conditions, light when added to light produces darkness. The phenomena belonging to this class cannot be explained, if following Newton, it is supposed that light is material. Two corpuscles coming together cannot destroy each other.

Another case considered by Newton was that of simultaneous reflection and refraction. To explain this he assumed that the particles had fits so that some were in a state favourable to reflection and others were in a condition suitable for transmission. No explanation of interference, diffraction and polarization was attempted because very little was known about these phenomena at the time of Newton. Further, the corpuscular theory has not given any plausible explanation about the origin of the force of repulsion or attraction in a direction normal to the surface.

Wave Theory

The test and completeness of any theory consists in its ability to explain the known experimental facts, with a minimum number of hypotheses. From this point of view, the corpuscular theory is above all prejudices and with its help rectilinear propagation, reflection and refraction could be explained.

By about the middle of the seventeenth century, while the corpuscular theory was accepted, the idea that light might be some sort of wave motion had begun to gain ground. In 1679, Christian Huygens proposed the wave theory of light. According to this, a luminous body is a source of disturbance in hypothetical medium called *ether*. This medium pervades all space. The disturbance from the source is propagated in the form of waves through space and the energy is distributed equally, in all directions. When these waves carrying energy are incident on the eye, the optic nerves are excited and the sensation of vision is produced. These vibrations in the hypothetical medium according to Huygens are similar to those produced in solids and liquids. They are of a mechanical nature. The hypothetical ether medium is attributed to the property of transmitting elastic waves, which we perceive as light. Huygens assumed these waves to be longitudinal, in which the vibration of the particles is parallel to the direction of propagation of the wave.

Assuming that energy is transmitted in the form of waves, Huygens could satisfactorily explain reflection, refraction and double refraction noticed in crystals like quartz or calcite. However, the phenomenon of polarization discovered by him could not be explained. It was difficult to conceive unsymmetrical behaviour of longitudinal waves about the axis of propagation. Rectilinear propagation of light also could not be explained on the basis of wave theory. The difficulties mentioned above were overcome, when Fresnel and Young suggested that light wave is transverse and not longitudinal as suggested by Huygens. In a transverse wave, the vibrations of the ether particles take place in a direction perpendicular to the direction of propagation. Fresnel could also explain successfully the rectilinear propagation of light by combining the effect of all the secondary waves starting from the different points of a primary wave front.

Huygens Principle

Huygens' principle gives a geometrical construction for finding the position of a wave front at a future instant if its position is known at some particular instant. The construction is based on the following two fundamental postulates.

(i) Every point on a wave front acts as a 'secondary' source of disturbance. Secondary wavelets spread in all directions from these new sources. The secondary wavelets are spherical and have the same frequency and velocity as the original wave.

(ii) The surface, which touches all the wavelets from the secondary sources, gives the new position of the wave front,

We now apply the Huygens' principle to the propagation of spherical wave fronts, of light from a point source, spreading in an isotro-pic medium.

Let, S be a point source of light producing spherical wave fronts. Suppose AB is the position of the wave front at some instant. According to Huygens' principle, every point A,..., C,...., E,.. .., B on AB is a source of secondary wavelets which advance with the velocity υ. Taking each of these points A,..., C,.... , E,...., B as a centre, spheres of radii equal to 'υt' are drawn.

These small spheres represent the secondary wavelets starting from these points. We now draw a surface MN which touches the small spheres in the forward direction. The tangential surface MN is the new wave front after an interval of t seconds. It is very easy to see that MN is the surface of a sphere having its centre at S.

Huygens' construction is an incomplete concept because it does not explain why there are no backward-going wavelets. In 1882 Kirchhoff eliminated this defect and showed that Huygens' primitive principle was a direct consequence of the differential wave equation. In spite of the defect, Huygens' principle served as very useful guide in explaining the phenomena of interference, diffraction and polarization.

Based on Huygens wave theory and Huygens principle, one can, by constructing the wavefronts, explain satisfactorily reflection and refraction of light. Adopting Fresnel's modification of Huygens principle, rectilinear propagation of light can also be explained.

Reflection of a Plane Wave Front at a Plane Surface

Let XY be a plane reflecting surface and AMB the incident plane wavefront. All the particles on AB will be vibrating in phase. Let be the angle of incidence (Fig. 8).

In the time the disturbance at A reaches C, the secondary waves from the point B must have travelled a distance BD equal to AC. With the point B as centre and radius equal to AC construct a sphere. From the point C, draw tangents CD and CD'. Then BD = BD'.

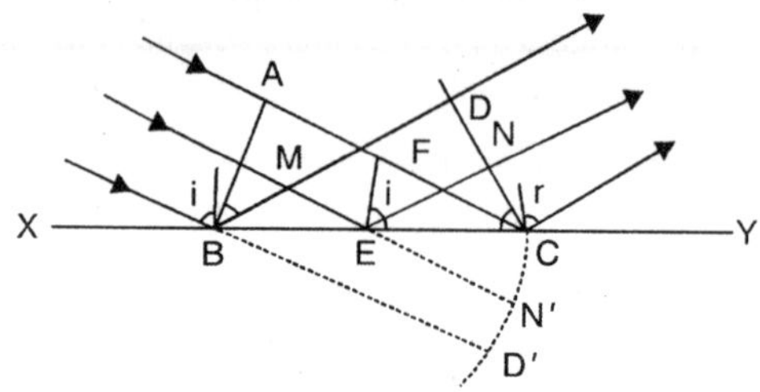

Fig. 8.

In the Δ^{les} BAC and BDC

BC is common, BD = AC and

$\angle BAC = \angle BDC = 90°$

$\therefore$ The two triangles are congruent.

$\therefore \angle ABC = i = \angle BCD = r$

$\therefore i = r$

Thus, the angle of incidence is equal to the angle of reflection.

ELECTROMAGNETIC THEORY

In 1862 Maxwell ingeniously synthesized electricity and magnetism and developed equations which succinctly combine the important theories. He showed that electromagnetic waves travel with the speed of light and hence drawn the most important conclusion that light wave itself is an electromagnetic wave. Initially, the existence of ether medium was presumed from propagation of electromagnetic waves in space. However, if light waves which are of very high frequency are to propagate, and at the same time allow a free passage to heavenly bodies, then the ether have to be rigid as well as pliable. It became impossible to visualize the hypothetical solid which could be easily compressed or extended, could permit resistance-free passage of heavenly bodies through it, and yet be elastic to twisting or bending stresses in order to allow propagation of waves. Ultimately, in 1887 Michelson and Morley proved conclusively that there was no ether surrounding the earth or elsewhere.

Quantum Theory

While experimenting on the black body radiation, Max Planck had come to the conclusion that the absorption or radiation of energy is not a continuous process. He postulated that thermal radiation is emitted or absorbed intermittently by indivisible amounts of energy called *quantum*. Each quantum carries an energy hv where h is a constant now called Planck's constant. Einstein elaborated the quantum concept in an endeavour to account for the phenomenon of photoelectric emission. He postulated that the quanta travel in space as separate entities with the speed 'c'. The quanta are named as photons. The further confirmation for the quantum theory is obtained when Compton effect was discovered in 1923. Compton found that when monochromatic x-rays fell upon matter, the scattered rays contained not only the original x-rays but also x-rays of wavelengths longer than the original. Though the quantum theory explains successfully the interaction of radiation with matter, it cannot account for the phenomenon of polarization, interference and diffraction. The contradictory aspects are reconciled by postulating dual nature of radiation. Accordingly, radiation is viewed as having both the particle as well as wave nature.

THE SOURCES OF LIGHT

The sun, the stars, lamps give off light. They are called luminous bodies. Other objects moon, mountains, trees etc. are non-luminous. They are visible only when they receive light from some luminous source and they send the light to our eyes.

Whether, a body is luminous or non-luminous depends on the conditions as well as on the material of which it is made. By changing the conditions we can make substances luminous or non-luminous. For *e.g.*, the filament inside the electric bulb is non-luminous unless it is heated by an electric current.

Bodies emit light at the expense of various kinds of energy. The most common is *thermal radiation*. When bodies are heated, to a temperature of 300°C they emit electromagnetic radiation, which lies in infrared region (λ, = 5 μm). They emit light as result of thermal motion of their molecules that is at the expense of their *internal energy*. At a temperature of 800°C, bodies emit visible radiant energy and appear red hot; a larger part of the energy still lies in IR region. At around 3000°C they appear white hot. Such heated materials are known as *incandescent* bodies. Not all sources are incandescent.

Some bodies can emit light, which is not due to transfer of thermal energy into the energy of electromagnetic waves. Emission of light due to supply of energy through processes other than heat is called *luminescence*. There are different kinds of luminescence. Advertisements using the neon and other glow tubes are examples of *electroluminescence*. In this kind of luminescence, charged particles accelerated by an electric field partly transmit their kinetic energy to the atoms of the gas, which then emit light. Many living organisms such as fireflies, fish and bacteria emit light due to chemical reactions. Such glow is called *chemiluminescence*. The cold light emitted by tube lights is a result of *phototuminescence*. The internal surface of the tube light is coated with a phosphor material, which under the action of UV light emits visible light. The TV screens and computer terminal screens glow because of *cathodo-luminescence*, which occurs due to bombardment of the screen by high-energy electrons.

REFRACTIVE INDEX

The refractive index of a medium is defined as the ratio of velocity of light in a vacuum to the velocity of light in the medium. Refractive index defined as above is called as *absolute refractive index*. Thus,

$$\mu = \frac{c}{\upsilon} \qquad \dots (1)$$

The refractive index is sometimes referred to as *optical density*. A medium with a relatively high refractive index is said to have a high optical density, while one with a lower index is said to have a low optical density.

The relative refractive index μ_{21} of a second medium relative to a first one is defined as the ratio of speeds of light υ_1 and υ_2 in the first and second media respectively.

The numerical value of refractive index is characteristic of the two media, but it also depends on wavelength. Hence, a refractive index is specified definitely only when wavelength is stated. Unless otherwise mentioned, refractive index is usually given for yellow light. The absolute refractive index for air under standard conditions is 1.0002918 for light having wavelength of the D-line of sodium (5893 Å). Since the refractive index for air is nearly equal to unity, it follows that for a solid or a liquid the absolute refractive index and the refractive index relative to air differ only slightly. Therefore, distinction is not made between them.

Note : The refractive index depends not only on the substance but also on the wavelength of the light. The dependence on wavelength is called *dispersion.*

OPTICAL PATH

The shortest distance, L between two points A and B is called the *geometric path.* The length of geometric path is independent of the medium that surrounds the path AB. When a light ray travels from the point A to point B, it travels with the velocity 'c' if the medium is air and with a lesser velocity v if the medium is other than air. Therefore, the light ray takes more time to go from A to B located in a medium.

From equation (1),

$$\mu = \frac{c}{\upsilon} = \frac{AN/t}{AB/T} = \frac{T}{t}$$

where r and T are the time taken by the light ray in air and in a medium respectively.

$$\therefore \qquad T = \mu t. \qquad ...(1)$$

The above relation means that a light ray takes μ times more time to cover the distance AB in a medium. To take into account the delay, we use another distance called the *optical length.* If a ray of light travels a distance L in a medium of refractive index μ in a certain interval of time, then it would travel a greater distance Δ in air during the same interval of time. Therefore,

$$\frac{\Delta}{L} = \frac{ct}{u\upsilon} = \mu$$

$$\text{or} \qquad \Delta = \mu L \qquad ...(2)$$

i.e., Optical path length = (Refractive index) (Geometric path length)

Thus, *the optical path length is defined as the product of refractive index and the geometric path length.*

Note that, if a ray travels a distance l in a medium of refractive index μ, the optical path length is equal to μl. In a given time light travels the same optical path length in different media. Suppose light travels a distance l_1 in a medium of refractive index μ_2 and a distance l_2 in a medium of refractive index μ_2 in time t. Then

$$\mu_1 l_1 = \mu_2 l_2.$$

RECTILINEAR PROPAGATION OF LIGHT

A straight line is the path of shortest distance between two points. Therefore, in a homogeneous medium the time taken by a ray of light along a straight line is a minimum as compared to any other path. It is obvious that the paths are straight lines, when light rays travel in a homogeneous medium.

REVERSIBILITY OF LIGHT RAYS

The optical path that is a minimum when light ray travels from point P to point Q is also a minimum when the ray travels in the opposite direction. Therefore, a ray that has traveled from P to Q will cover the same path in the opposite direction from Q to P. Thus, the reversibility of light rays follows from Fermat's principle.

LAWS OF REFLECTION

We wish to find the path that a light ray takes in going from A to C after first striking the mirror MN such that the time taken is a minimum.

One way would be to go to the mirror quickly and then to go to C along the path ADC. Though the path AD is short, the path DC is longer. In order to find the path, for which the time would be shortest, a geometrical trick is used. Point I is located behind the mirror, such that CF = IF. In the Δ^{10} CDI, $\angle CFD = 90°$ CF = FI and DC = DI. Therefore, the sum of the distances AD + DC, which is proportional to the time taken for light, is also the sum of the two lengths AD + DI. Then the problem consists in finding the path that a ray from A takes in going to I in the shortest time. Is it path ADI? As AD and ID are two sides of the triangle AID, their sum is greater than the length of the third side AI. So the answer is no. Perhaps it is the path AEI? The answer is again no for exactly the same reason: AE + EI > AI. Proceeding in this way we arrive at the only possible answer the ray must travel along path ABI because this is the shortest route. This means that in going from A to C, the ray is reflected at B. Let us suppose that a ray from A strike the mirror at 0 before being reflected to B. The basic method is, first, to obtain an expression for the path length L traveled by the ray from A to B, in terms of x, *l*, a and b. Secondly, minimize L by altering the position of point O. Thirdly, substitute the angles; and r into the resulting expression. The minimum path length may now be found by differentiating L with respect to x and equating the resulting expression to zero.

First Law of Reflection

Consider a plane mirror. A and B are two points above the mirror plane and are located in a plane ABCD normal to the mirror plane. Light coming from point A is reflected toward B (Fig. 2.7). Suppose the light ray passes through a point P. It means that the light is incident along AP and is reflected along PB. Thus, APB is a most general conceivable path from A to B.

Draw a plane ABCD normal to the mirror plane $M_1M_2M_3$. Take the origin of coordinates at D. Let DC and DA be the x and y axes. Let DA = a, CB = b and DC = d. The point P has general coordinates (x, 0, z). If now AP + PB = L, we get

$$L = [(x-0)^2 + (0-a)^2 + (z-0)^2]^{1/2} + [(x-d)^2 + (0-b)^2 + (z-0)2]^{1/2}$$

$$\text{or} \quad L = \sqrt{x^2 + a^2 + z^2} + \sqrt{(x-d)^2 + b^2 + z^2} \qquad ...(1)$$

We now apply Fermat's principle to get the actual path. The path APB can be varied by varying x and z. We obtain the minimum value of L, that is the shortest optical path by taking the derivative of L with respect to z and setting the derivative equal to zero. Thus,

$$\left(\frac{\partial L}{\partial z}\right)_x = \frac{1}{2}\left[\frac{1}{\sqrt{x^2+a^2+z^2}}.2z\right] + \frac{1}{2}\left[\frac{1}{\sqrt{(x-d)^2+b^2+z^2}}.2z\right] = 0$$

$$\therefore \quad z\left[\frac{1}{\sqrt{x^2+a^2+z^2}} + \frac{1}{\sqrt{(x-d)^2+b^2+z^2}}\right] = 0$$

As the factor within the brackets cannot be zero, z must be equal to zero.

$$\therefore \qquad z = 0$$

The above result means that P must lie in the plane ABCD, which is normal to the mirror. O is such a position for P. If P coincides with O, it is obvious that the incident ray AO, the surface normal ON and the reflected ray OB lie in the same plane. This is the first law of reflection.

Second Law of Reflection

Using z = 0 into eqn. (1), we get

$$L = L = \sqrt{x^2 + a^2} + \sqrt{(x-d)^2 + b^2} \qquad ...(2)$$

Now taking the derivative of L with respect to x, we get

$$\frac{dL}{dx} = \frac{1}{2}\frac{1}{\sqrt{x^2 + a^2}}.2x + \frac{1}{2} + \frac{1}{\sqrt{(x-d)^2 + b^2}}.2(x-d)$$

Setting dL/dx = 0, we obtain

$$\frac{x}{\sqrt{x^2 + a^2}} + \frac{(x-d)}{\sqrt{(x-d)^2 + b^2}} 0$$

$$\therefore \quad \frac{x}{\sqrt{x^2 + a^2}} = \frac{-(x-d)}{\sqrt{(x-d)^2 + b^2}}$$

or $$\frac{x}{\sqrt{x^2 + a^2}} = \frac{(d-x)}{\sqrt{(x-d)^2 + b^2}}$$

Looking at triangles AOD and BOC

$$\sin i = \frac{x}{\sqrt{x^2 + a^2}} \text{ and } \sin r = \frac{(d-x)}{\sqrt{(x-d)^2 + b^2}}$$

$$\therefore \quad \sin i = \sin r$$

$$\therefore \quad i = r \qquad ...(3)$$

This is the second law of reflection.

We now apply Fermat's principle for refraction phenomenon. We have a boundary PQ between two media. Light goes from A to C. If the speed of light on both sides of the boundary were the same, the path from A to C would be a straight line. The speed above PQ and that below PQ are different and we assume that the medium above PQ is air and below PQ is water.

Feynman has given an apt analogy which we cite here. "To illustrate that the best thing to do is not just to go in a straight line, let us imagine that a beautiful girl has fallen out of a boat, and she is screaming for help in the water at point C. We are at point A on land and we see the accident, and we can run and can also swim. But we can run faster than we can swim. What do we do? Do we go in a straight line? By using a little intelligence we would realize that it would be advantageous to travel a greater distance on land in order to decrease the distance in the water, because we go so much slower in the water."

Consider a plane surface S separating two media. Let A and C be two points lying in the two different media. We must find the path from A to C, which can be covered by light faster than it can cover any other hypothetical path. Clearly, the path must consist of two straight lines namely AB in medium 1 and BC in medium 2; the point B in the plane S has to be found.

Let K be any general point on the surface S. It has the coordinates (x, 0, z). The optical path through the point K (x, 0, z) is given by

$$\Delta = (AK)\,\mu_1 + (KC)\,\mu_2$$

$$= \mu_1\,(x^2 + a^2 + z^2)^{1/2} + \mu_2\,[(x - c)^2 + b^2 + z^2]^{1/2} \qquad ...(4)$$

Now apply Fermat's condition to get the actual path.

$$\left(\frac{\partial \Delta}{\partial z}\right)_x = 0$$

$$\therefore \qquad z = 0$$

It means that the incident ray, the refracted ray and the normal to the plane EQRS lie in the same plane. This is the first law of refraction.

Applying the second condition that $\left(\frac{\partial \Delta}{\partial z}\right)_x = 0$, we get,

$$\left(\frac{\partial \Delta}{\partial z}\right)_x = \frac{\mu_1}{2}\left[\frac{2x}{\sqrt{x^2 + a^2 + z^2}}\right] + \frac{\mu^2}{2}\left[\frac{2(x-c)}{\sqrt{(x-c)^2 + b^2 + z^2}}\right] = 0$$

$$\frac{\mu_1 x}{\sqrt{x^2 + a^2 + z^2}} = \frac{\mu^2(c-x)}{\sqrt{(x-c)^2 + b^2 + z^2}}$$

$$\mu_1 \sin i = \mu_2 \sin r$$

$$\therefore \qquad \frac{\sin i}{\sin r} = \frac{\mu_2}{\mu_1} \qquad ...(5)$$

OPTICAL PATH STATIONARY

Let us consider reflection from the inner surface of an ellipsoid of revolution. F_1 and F_2 are the foci of the ellipsoid. According to the definition of an ellipse, the paths F_1OF_2, F_1, O' F_2, F_1, O" F_2 etc. are identical in length. Hence all the rays leaving focus F_1 and arriving after reflection at focus F_2 take same time. In this case, the optical path is stationary. It is also a geometrical property of the ellipse that i = r for any location of O. All optical paths from F_1 to F_2 via a reflection are therefore precisely equal. None is a maximum. The optical path length is clearly stationary with respect to variations. If we replace the surface of the ellipsoid with surface MN having a smaller curvature and oriented so that a ray leaving point F_1 arrives at point F_2 after being reflected from MN, then the path F_1OF_1 will be a minimum. Let MN be a curved mirror. Let F_1OF_2 be the actual path of a ray from F_1 reaching F_2 after reflection at this mirror. Let M' be a point on the mirror near O, so that $F_1M'F_2$ be a neighbouring path. Let F_1M' intercept the ellipse at M".

$F_1O + OF_2 = F_1M'' + M''F_2$ = constant (Property of the ellipse)

In the neighbouring path $F_1M'F_2$ is given by

$$F_1M' = F_1M'' + M''M'$$

The difference between the actual path F_1OF_1 and the neighbouring path $F_1M'F_1$ is Δ.

$$\begin{aligned}\Delta &= F_1OF_2 - F_1M'F_2\\ &= (F_1O + OF_2) - (F_1M' + M'F_2)\\ &= F_1M'' + M''F_2 - F_1M'' - M''M' - M'F_2\\ &= M''F_2 - (M''M' + M'F_2) \qquad \text{...(1)}\end{aligned}$$

This quantity is always negative. In triangle $M'F_2M''$

$$M''M' + M'F_2 > M''F_2$$

∴ The actual path is the minimum (shortest) among the neighbouring paths.

In case of the surface NN' having a curvature greater than that of ellipsoid, path F_1OF_2 will be a maximum. Now, we have to extend F_1M' to meet the ellipse at M".

Now $\qquad F_1O + OF_2 = F_1M'' + M''F_2$

The difference between the actual path F_1OF_2 and the neighbouring path $F_1M'F_2$ is Δ.

$$\Delta = F_1OF_2 - F_1M'\,F_2$$
$$= (F_1O + OF_2) - (F_1M' + M'F_2)$$
$$= F_1M'' + M''F_1 - (F_1M'' - M''M') + M'\,F_2$$
$$= M''F_2 - (M'\,F_2 - M'\,M'') \quad ...(2)$$

which is always positive. The actual path is the *maximum* (longest) among the neighbouring paths.

Obviously, the optical path is stationary among the neighbouring paths, if the reflecting surface coincides with the locus surface

$F_1O + OF_2$ = constant.

THE CONCEPT OF RAYS OF LIGHT

Among the simplest laws concerning the travel of light which we will examine are the following:

(a) Law of mutual independence of rays of light passing through a given medium.

(b) Law of reversibility path of light.

If several beams of light are passing through a given medium simultaneously in different directions, then the path of any beam is the same as it would be if all other beams were absent. That is the mutual independence referred to in law (a). The law of reversibility of paths means that if a ray AB follows a course ABCDE, then if we start with a ray ED, it must follow the path EDCBA.

All such laws are good and valid ordinarily. The basic assumptions involved in all these laws are that light energy travels along a well defined path, which we call a ray. But when we take a closer look at these assumptions we face problems. A beam of light starting from 0, spreading to width PQ, and then passing to the second 1 medium on the right. In the figure we divide this beam into 5 parts, which are numbered 1, 2,... 5. You may imagine the beam divided into 5,000 parts in this way, or even 5×10^{20} parts if you like to get to your 'ray.' But you cannot ignore the fact that the beam spreads as it travels. If this is the situation, what happens to the law of reversibility? The middle line (shown by dashes) in the left and right diagrams would match, but the spreads would not match. The spread speaks of the time sequence A to E for the first diagram and E to A for the second diagram. Let us now have a look at the law of *mutual independence* of rays. In Fig. 2.3 let the beam be divided into, say, 100 parts 1 to 100. Then the path followed by any one beam (say beam 50) when it is alone is not the same as the path when neighbouring beams are also, present.

The result. If ray number 50 alone is allowed to go ahead it spreads out, overand above the spread which occurred earlier. Thus neighbouring rays do have influence on the course which any given ray is to follow; the influence is of reducing any increase in the spread.

So long as the apertures of the beams are not too narrow, these effects are not serious, and we continue to treat the simple laws as applicable. We may define 'Geometrical Optics' as: that branch of optics in which the geometrical laws of rectilinear propagation, reflection and refraction of light are good enough approximations to physical reality.

In the rest of this chapter we shall discuss a principle of very wide-ranging importance in so far as it unifies in itself all the laws of geometrical optics, and at the same time also builds a bridge between geometrical and physical optics. This is Format's principle.

It is found that there are a number of cases in which the real path of light is the one for which the time taken is not a minimum but a maximum. For example, in case of a spherical reflector, light prefers maximum time. On the other hand, in elliptical reflectors light ray takes the same time for all paths. In view of these facts the Format's principle is to be modified. In its modem form the principle states that,

A light ray travelling from one point to another point will traverse a path for which, compared to all neighbouring paths, the time required is a minimum or a maximum or stationary.

This is known as *"Format's principle of extremum path"* or *"Format's principle of stationary time"*.

By a stationary value, we mean one for which dL/dx has a maximum, or a minimum or a point of inflection with a horizontal tangent. Curve of the OPL versus x will have a somewhat flattened region in the vicinity of where the slope goes to zero. The zero-slope point corresponds to the actual path taken. The OPL curve will look like in case of refraction. Here OPL is a minimum. A small change in x in the vicinity of O has little effect on the OPL, but a similar change in x anywhere well away from O results in a substantial change in OPL. Light takes a path such that there are many other paths nearby which take almost exactly the same time.

The fundamental laws of rectilinear propagation, reflection and refraction can be derived from Fermat's principle.

PARABOLIC MIRROR

Let RS represent a parabolic mirror. It can be shown that rays incident on the mirror parallel to its axis are reflected through its focus. Let us consider the ray path be PQ' P, which is not parallel to the axis of the mirror. According to Fermat's principle the ray path will correspond to a minimum. From the point Q' we drop a perpendicular Q' N' on the directrix AB. From the definition of parabola it follows that Q' N' = Q' F.

Thus, PQ' + Q'F = PQ' + Q'N'

Let N be the foot of the perpendicular drawn from point P on AB. Then for PQ' + Q' N' to be a minimum, the point Q' should lie on the straight line PN. The actual ray will be PQ + QF where PQ is parallel to the axis. Therefore, all rays parallel to the axis will pass through F and conversely all rays emanating from the point F will become parallel to the axis after reflection.

CHROMATIC DISPERSION

The refractive index is a function of wavelength. Generally, it decreases as the wavelength increases. Consequently, light passing through a material medium will be separated according to wavelength. This is known as chromatic dispersion.

With the light at normal incidence, blue light would merely lag behind the red light. With the light incident obliquely, the colours become separated in space, both surfaces refracting blue more than red. With a prism the separation is even more pronounced.

Note that dispersion is not proportional to refraction. Refraction and dispersion bear no simple relationship to one another. Some glasses have a high index of refraction and little dispersion; others have just the opposite.

(a)

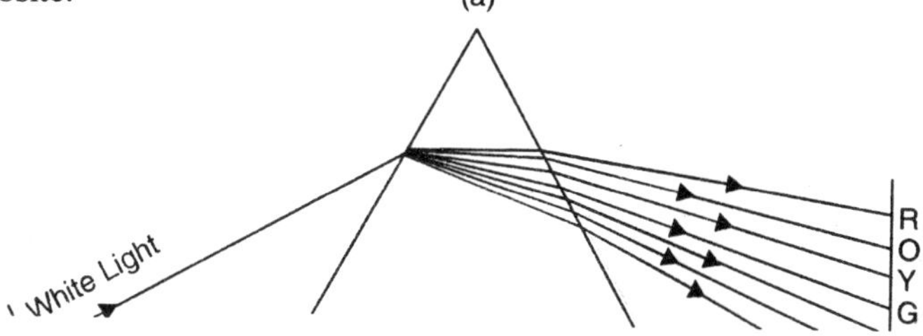

Fig. 9.

When a ray of white light falls on a glass prism, it splits up into different colours (see Fig. 9). This display of colours is known as *spectrum* of the source of light. This separation of a composite beam into its constituent colours is called *dispersion* and the medium which produces dispersion is known as a *dispersive medium*. When the colours travel through a refracting medium, their velocities are different. Hence, the refractive index of the material is different. Out of the seven colours formed by the white light, violet colour suffers greater deviation than red light.

THE VELOCITY OF LIGHT

The first attempt to measure the velocity of light was made by Galileo in 1667. Two observers were stationed on the tops of two hills about 1.5 km apart. Each observer was provided with one lamp and the experiment was carried out at night. One observer, say A, uncovered his lamp thereby sending a short flash of light to the other observer, say B and noted the instant at which he did so. The other observer B uncovered his lamp as soon as he saw the flash from B's lamp. The interval between these two instants of time noted by A is obviously equal to the time taken by the light to travel the distance AB twice. However, the interval could not be found between the two instants. From this Galileo concluded that velocity of light, if finite, was extremely large.

Eight years later, in 1675, the Danish astronomer, Roemer succeeded in measuring the velocity of light. The satellites of planet Jupiter show eclipses once in every 7 days. He noticed that the time interval between two consecutive eclipses of one of the satellite of Jupiter varied with the relative positions of the earth. This time interval increased as the earth moved away from Jupiter and decreased as it moved toward Jupiter. Roemer correctly attributed this variation to the finite velocity of light and found that light requited about 22 minutes to travel a distance equal to the diameter of earth's orbit; The diameter of the earth's orbit was known to be 2.87×10^{11} m.

$$\therefore \qquad c = \frac{2 \times 10^{11}\text{m}}{22 \times 60\text{s}} = 2.3 \times 10^{8}\ \text{m}.$$

This value is lower than the action speed. Regardless of its accuracy it showed that the speed of light was not infinite, as many then believed, but was finite and measurable.

The first laboratory measurement of the speed of light was carried out in 1849 by the French physicist H.L. Fizeau. Later, many experimenters using different methods have measured c. The most precise determination gives a value,

$$\therefore \qquad c = 2.997924 \times 10^8 \text{ m/s.}$$

which is usually simplified as

$$c = 3 \times 10^8 \text{ m/s.} \qquad ...(6)$$

VISIBLE RANGE

The arrangement of the various electromagnetic waves in a continuous sequence of frequencies and wavelengths is called an *electromagnetic spectrum*. The spectrum includes waves covering a broad range of wavelengths. It is bounded at one end by the gigantic radio waves having wavelengths of a few kilometres and at the other end by y- rays of tiny wavelengths of the order of 10^{-12} m. *Visible range* is that part of the spectrum constituted by waves, which can be detected by the human eye. It extends from the deepest violet to the deepest red. The limiting range of these waves depends in the individual properties of the eye and varies approximately in the interval λ, = 4000Å to λ = 7800Å. The regions flanking the visible range are *infrared (IR)* on the longer wavelength side and *ultraviolet (UV)* on the shorter wavelength side. The IR region lies in the wavelength range 7.8×10^7m to 10^3m and the UV region lies between 4000Å to 10Å. The radiation in these three regions, namely visible, IR and UV, put together is called *optical radiation*. Table 1 lists the wavelengths and the frequencies of each of the coloured region in the visible region.

The sensitivity of the human eye is a function of wavelength. It has peak *sensitivity* at a wavelength of about 5500 A corresponding to yellow-green.

SOLVED EXAMPLES

Example 1:

The contact faces are both plane, complete the 'design' of the doublet.

Solution:

$R_1' = R_2 = \infty$. Hence,

$$\frac{1}{R_1} = G_1, \text{ and } -\frac{1}{R_2'} = G_2$$

Using the values of G_1 and G_2 deduced above, we get

$R_1 = 1711$ cm, $R_2' = 28.4$ cm.

Example 2(a):

Calculate the dispersive power for crown and flint glass from the following data:

	C	*D*	*F*
Crown	*1.5145*	*1.5170*	*1.5230*
Flint	*1.6444 1.*	*6520*	*1.6637.*

Solution:

$$\omega_1 = \frac{\mu_F - \mu_C}{\mu_D - 1} = \frac{1.5230 - 1.5145}{1.5170 - 1} = 0.01644$$

$$\omega_2 = \frac{\mu_F - \mu_C}{\mu_D - 1} = \frac{1.6637 - 1.6444}{1.6520 - 1} = 0.02961$$

Example 2(b):

For a crown and flint glass for C and F lines $\mu_C = 1.515$, and $\mu_l = 1.523$ and $\mu_C = 1.644$, $\mu_F = 1.664$ respectively. Calculate the angle of flint glass prism which may be combined with crown glass prism having refracting angle 20° so that the combination is achromatic for C and F rays.

Solution:

The condition for the combination to be achromatic is that

$$(\mu_F - \mu_C)\,A = (\mu_F' - \mu_C')\,A'$$

$$(1.523 - 1.515) \times 20° \; (1.664 - 1.644)\,\alpha'$$

$$\alpha' = \frac{0.008}{0.02} \times 20° = \mathbf{8°}.$$

Example 2(c):

Reverting to the situation of example 1, let 70 messengers start simultaneously from A to reach B through paths with s = 0, 1, 2, ... 70. If the speeds of all are equal, being 10 speed units, deduce from the plot of t versus s how many of the messengers reach within 0.1 time unit of the arrival of the messenger reaching earliest.

Solution:

Converting p/speed = time t, the data from solution of example 1 become the following:

s = 0	10	20	30	40	50	60	70
p =11.61	10.97	10.68	10.64	10.83	11.21	11.81	12.62

The graph (not shown) gives minimum time = 10.62 units (for messenger number 26). Within 0.1 unit of time 10 other messengers reach B (messenger numbers 17 to 35). [Note that in the next 0.1 unit time interval only 5 or 6 messengers would reach].

Example 2(d):

Let A and B be two points in a plane. P is a variable point defined by CP = s, and AP + PB is a variable path length p dependent on s. Take s = 0, 10, 20, ... 70 units, compute p each time, and draw a graph for p versus s.
Find if there is a path which is maximum or minimum amongst neighbouring paths.

Solution:

From geometry of the figure

$$p = [30^2 + s^2]^{1/2} + [50^2 + (70 - s)^2]^{1/2}$$

Straight computation gives the following:

s = 0	10	20	30	40	50	60	70
p = 116.1	109.7	106.8	106.4	108.3	112.1	118.1	126.2

The graph between p and s may now be drawn. We get a *minimum* p at s = 26.

Example 3:

Suppose the messengers have to pick some load from positions s so that speed over distance PB is reduced to half, i.e., 5 speed units. Which messenger now will have the shortest timing?

Solution:

$$\text{Now } t = \frac{1}{10}[30^2 + s^2]^{1/2} + \frac{1}{5}[50^2 + (70-s)^2]^{1/2}$$

The messenger with least time among neighbours is now number 47 (from graph) with timing 16.59 units.

The purpose of the sequence of examples given above is to show the meaning of Fermat's principle in depth. The 'messengers' of the light signals starting out from A. The real signals reaching B are those which have crowding of neighbouring signals in time. This condition is satisfied at dt/ds = 0. One may check also that in example 1 for s = 26 the angles which AP and PB make with the normal to CD are equal, and in Example 1.3 for s = 47 the ratio of sines of these angles is 2, equal to the ratio of the speeds.

Fermat's principle in effect tells us that no ray is to be considered in isolation. It specifically asks us to extend our examination to the neighbourhood of the considered path (or ray), to check if timings for those paths would *crowd together* around that for the path under consideration. If we can establish the answer as 'yes' then alone the path considered becomes a real path.

Example 4:

Dispersive powers of Crown and Flint glasses are .0163 and 0243 respectively Design an achromatic contact doublet of focal length 50cm.

Solution:

Straight substitution in Eqns. (12) and (13) gives

$$\frac{.0163}{f_1'}+\frac{.0243}{f_2'}=0;\quad \frac{1}{f_1'}+\frac{1}{f_2'}=\frac{1}{50}$$

Solving, we get $f_1' = \dfrac{4000}{243} = 16.5$ cm (*crown*)

$$f_2' = \frac{4000}{-163} = -24.5 \text{ cm } (\textit{flint}).$$

Example 5:

A contact achromat of power + 1 is to be made using crown and flint glass lenses. For crown n_C = 1.51645, n_F = 1.52492. For flint n_C = 1.57446, n_F = 1.58854. Calculate the focal lengths of the component lenses.

Solution:

Nature of data desire using Eqns. (11) and (9). Now,

using suffixed 1 and 2 for frown and flint, respectively, we have

$n_{1F} - n_{1C} = 0.00847,\ n_{2F} - n_{2C} = 0.01408$

For achromatism, Eq. (11) gives

$0.00847\ G_1 + 0.01408\ G_2 = 0$

For the combination power + 1, f ' = 100 cm. Hence from Eq. (9).

$$0.5164\ G_1 + 0.5745\ G_2 = \frac{1}{100}$$

We have to solve these two equations for G_1 and G_2. This gives

$G_1 = 0.0585\ G_2 = 0.0352$ (check it)

Finally,

$$\frac{1}{f_i C} = 0.5164\ G_1 = 0.5164 x\ 0.0585$$

$\therefore \qquad f_{1C} = 33.1$ cm (*crown*)

Similarly, $\quad f'_{2C} = -\ 49.4$ cm (*flint*).

Example 6:

A point source distant 0.6 from the centre C of a spherical reflecting surface of radius r. From Fermat's Law show that if ray PA is reflected to Q, the angle θ is given by $\cos^2\theta/2 = 2/3$.

Solution:

In general, let PC = a.r.

Then $p = PA + AQ = r\ (1 + a^2 - 2a \cos \theta)1/2 + 2r \cos (\theta/2)$

Applying Fermat's law $\partial p/\partial\theta = 0$ leads to two solutions

$$\sin \theta/2 = 0;\ 2a \cos \theta/2 = (1 + a^2 - 2a \cos \theta^{1/2}).$$

The first gives one possible incident ray as PQ', which after reflection passes through Q. The second one leads to

$$\cos^2 (\theta/2) = \frac{1+a}{4a}$$

That gives the desired result for a = 0.6.

[*Note:* If a < 1/3, the only raw from P which can reach Q is PQ'.]

Example 7:

C is the centre of a refracting sphere of radius r, refractive index n = 1.50. A parallel beam is incident in the direction OC. Find out that ray which after refraction passes through a point Q distant CQ = 1.2r.

Solution:

For all incident rays the path up to the dotted line OA' is common. So the variable path is p = A'A + nAQ. The parameter convenient to take is $\angle OCA = \theta$.

$$p = A'A + nAQ = r\,(1 - \cos\theta) + nr\,(1 + a^2 + 2a\cos\theta)^{1/2}$$

Applying Fermat's law $\partial p/\partial\theta = 0$ leads to two solutions:

either' $\sin\theta = 0$

or $(1 + a^2 + 2a\cos\theta)^{1/2} = na$

The first solution gives the ray incident along OC, which passes through all positions of Q on line OC. The second gives

$$\cos\theta = \frac{(n^2 - 1)a^2 - 1}{2a}$$

In the present case n = 1.50, a = 1.2

$\therefore \quad \cos\theta = \frac{1}{3}$ (check it)

Note: For $\theta \to 0$, leads to $a = (n-1)^{-1}$ which gives a = 2 for the given n.

So, you may see the "spherical aberration". 0.8a as θ goes from 0 to $\cos^{-1}\left(\frac{1}{3}\right)$.

Example 8:

Aplanatic points case. In a sphere of radius r and refractive index n surrounded by air, consider points A and B situated at distance r/n and rb respectively from the centre along a given radial line. Deduce that ray from A which, after refraction, appears to come from B.

Solution:

OA = r/n and OB = rb. Now consider a point P on the sphere specified by parameter θ. We join AP, and join BP and produce it to B'. The actual conceivable path of a ray is APB. But for a path "from A to B via P" the path will be treated as AP in the medium *minus* PB in air. Whether APB' is or is not a real path only Fermat's principle will tell us.

So we proceed to compute the optical APB = p.

$$p = nAP - PB$$

$$= n\left[\frac{r^2}{n^2} + r^2 + \frac{2r^2}{n}\cos\theta\right] - [r^2b^2 + r^2 + 2r^2b\cos\theta]$$

Applying Fermat's law $dp/d\theta = 0$ leads to one condition $\sin\theta = 0$, which is trivial. The other condition, on simplification, comes to

$$\frac{1}{n^2} + \frac{2}{n}\cos\theta = \frac{1}{b^2} + \frac{2}{b}\cos\theta$$

Now, for a general value of b this condition is not satisfied for any value of θ. This is a new situation. Now ray from A, whatever its point of incidence P at the sphere, can pass through the point specified by OB = rb for a general value of b.

If, however, the point B is specified by OB = nr (*i.e.*, we have b = n). But again there is a new situation is now satisfied for *all values* of θ. Hence, not only a particular ray but *all the rays* from *A will, after refraction, have B on their backward projection.*

Example 9:

A crown glass prism of refracting angle 8° is combined with a flint glass prism to obtain deviation without dispersion. If the refractive indices for red and violet rays for crown glass are 1.514 and 1.524 and for the flint glass are 1.645 and 1.665 respectively, find the angle of flint glass prime and net deviation.

Solution:

The condition for deviation without dispersion is

$$(\mu_V - \mu_R)\,A = (\mu_V' - \mu_R')\,A'$$

$$\therefore\quad A' = \frac{(1.524 - 1.514)\times 8^\circ}{(1.665 - 1.645)}$$

$$= \frac{0.08^\circ}{0.02} = 4^\circ$$

Mean refractive index for crown glass $\mu = \dfrac{1.514 + 1.524}{2} = 1.519$

Mean refractive index for flint glass $\mu = \dfrac{1.645 + 1.665}{2} = 1.655$

$\therefore$ The net deviation $(\delta - \delta) = (\mu - 1)\,A - (\mu' - 1)\,A'$

$= 0.519 \times 8^\circ - 0.655 \times 4^\circ$

$= 1.53^\circ.$

Example 10:

A lens made of crown glass has focal length 50 cm. Calculate the longitudinal chromatic aberration given, n_C = 1.5164, n_F = 1.5249.

Solution:

n_D being not given, we use mean n = 1.5206 for it.

$$\omega = \frac{1.5249 - 1.5164}{0.5206} = \frac{.0085}{.5206} = 0.0164$$

Longitudinal chromatic aberration $x = f'\,\omega = 50 \times .0164 = 0.82$ cm.

Example 11:

If aperture of the eye is taken as 2 mm and least distance of distinct vision is 25 cm, calculate the 'numerical aperture' of the unaided eye. The best numerical aperture obtained in a microscope approaches a value 2. How much should the normal magnification of such microscope be?

Solution:

Semiangle = 0.1 cm/25 cm = 0.004 radian. For air n = 1. Hence.

N.A. of the eye = 0.004 radian

Considering 1.22 λ term common between eye and a microscope, the detail resolved by the objective O the microscope is smaller by factor

$$\frac{\text{N.A.of microscope}}{\text{N.A.of the eye}} = \frac{2}{.004} = 500$$

So magnification should be > 500 fold if the microscope is to use full resolution. 'Normal' magnification = 500.

Example 12(a):

Let the focal length of an eyepiece be 2.5 cm. For a telescope of magnification 100, deduce the focal length of the objective and useful diameter of the objective, given d_{eye} = 2 mm.

Solution:

Given $m = \frac{f_0'}{f_E'} = 100$ and $f_E' = 2.5$ cm

$\therefore \quad f_0' = 250$ cm

For optimum use, $\frac{d_0}{d_E} = m;$

and $d_E = 2$ mm (given)

$$d_0 = 20 \text{ cm.}$$

Example 12(b):

In the case of Example 6.1 add a lens of f' = 12 cm at the position of the primary image, giving it a 'suitable' aperture to use the full aperture of the eyelens. Calculate (i) aperture of this field lens, (ii) new semi-angle of the field of view.

Solution:

With O as centre of the objective, OBE is the path of marginal rays when L_2 alone is present. With L_1 added, the path of marginal rays becomes OABE'. Here lens L_1 is the field lens added, making $L_1 + L_2$ the two-lens eyepiece. We have

$$OL_1 = 100 \text{ cm}, \; L_1L_2 = 6 \text{ cm}$$

whence we deduce (check them);

$$L_1P = 13.6 \text{ cm}$$

$$\Rightarrow \qquad L_2P = 7.6 \text{ cm}$$

$$\text{aperture of} \qquad L_1 = 3 \times \frac{13.6}{7.6} = 5.3\text{cm}$$

$$\text{now} \qquad \theta_1 = \frac{1}{2} \text{x} \frac{5.3}{100} = .0265 \text{ radian.}$$

Thus the field of view increases to about 2-fold. You may note that lens L1 here does not play a role in image formation (because its location coincides with the primary image); yet it enhances the field of view. The eye ring shifts from E to E', so θ_2 also increases by the same factor as θ_1, keeping the magnification unchanged.

Example 12(c):

Find the focal length, unit planes of Huygens's eye piece.

Solution:

Huygens's eye piece consists of two Plano convex lenses each of focal lengths 3f and f separated by a distance of 2f. The Huygens's eye piece.

The system matrix for the lens $L_1 = \begin{pmatrix} 1 & \frac{-1}{3f} \\ 0 & 1 \end{pmatrix}$

The translation matrix from L_1 to $L_2 = \begin{pmatrix} 1 & 0 \\ 2f & 1 \end{pmatrix}$

The system matrix for the second lens $= \begin{pmatrix} 1 & \frac{-1}{f} \\ 0 & 1 \end{pmatrix}$

The system matrix of Ramsden's eye piece

$$S = \begin{pmatrix} 1 & \frac{-1}{f} \\ 0 & 1 \end{pmatrix} \begin{pmatrix} 1 & 0 \\ 2f & 1 \end{pmatrix} \begin{pmatrix} 1 & \frac{-1}{3f} \\ 0 & 1 \end{pmatrix} = \begin{pmatrix} 1 & \frac{-2}{3f} \\ 2f & \frac{1}{3} \end{pmatrix}$$

Comparing this with the general system matrix, Gaussian constants are

$$a = \frac{2}{3f};\ b = -1;$$

$$c = \frac{1}{3};\ d = -2f$$

The combined focal length of the system F is given by

$$F = \frac{1}{a} = \frac{3f}{2}$$

The distances of the unit planes

$$\alpha = \frac{1-b}{a} = \frac{1+1}{2} x 3f = 3f$$

$$\beta = \frac{c-1}{a} = \frac{(1/3)-1}{2} x 3f = f$$

Nodal planes coincide with unit planes.

Example 13:

Obtain the system matrix for a thin lens placed in air and made of refractive index 1.5 and radii of curvature 50 cm each.

Solution:

System matrix for a thick lens is given by

$$S = \begin{pmatrix} 1 & -P_1 - P_2 \\ 0 & 1 \end{pmatrix}$$

The powers of two surfaces are equal. Hence

$$P_1 = P_2 = \left(\frac{\mu - 1}{R_1}\right)$$

$$= \left(\frac{\mu - 1}{|R_2|}\right) = \frac{1.5 - 1}{50} = 0.01$$

$$\therefore \quad S = \begin{pmatrix} 1 & -0.01 - 0.01 \\ 0 & 1 \end{pmatrix}$$

$$= \begin{pmatrix} 1 & -0.02 \\ 0 & 1 \end{pmatrix}$$

Further $\qquad f = \dfrac{1}{0.02} = 50 \text{ cm.}$

Example 14:

The lens of Example 7 is used to form the image of a point source placed 70 cm in front. Deduce the longitudinal focussing difference for C and F light.

Solution:

One can easily get v = 175 cm. Then using Eq. (7), and w = 0.0164 obtained previously,

$$\nabla v = .0164 x \frac{(175)^2}{50} = 10.0 \text{ cm.}$$

(Notice the largeness of the error. In daylight we do not have much intensity in F region. So the observed aberration is between C and D light, which is smaller).

Example 15:

Consider Foucault's experiment, using 14 transits over a distance D arranged by using mirrors causing forward and backward reflections. If speed of rotation of the mirror is 10 rev/sec., and focal length of the lens system is 50 cm, deduce the D needed to get a mm displacement of the final image.

Solution:

$$\Delta t = \frac{14\times 1)(\text{in metres})}{3\times 10^8\,\text{m/sec}} = \frac{141}{3}10^{-8}5$$

rotation of the mirror, $\alpha = 2\pi n\ \Delta t$ radian $= \frac{28\pi D}{3}\times 10^{-7}$ radian

Shift of image 0.1cm $= 2\pi f = \frac{28\pi D}{3}\times 10^{-5} \Rightarrow D = 3.8$ m.

Example 16:

In a playground we arrange a short flash at one end, and a mirror at the opposite end 100 metres away. The original flash and the return flash are both fed to fast time-recorder (like a cathode-ray oscilloscope). What difference of time would it record?

Solution:

$$\nabla t = \frac{2\times 100\,\text{m}}{3\times 10^8\,\text{m/sec}} = 0.67 \times 10^{-6} \text{ sec.}$$

[Note: Oscilloscopes with 1 megacycle (10^6Hz) frequency are now easily available. Their screen divided into 100 parts gives a least account of. 01 $\times 10^{-6}$ sec. The problem really is of producing a flash of duration $< 10^{-6}$ sec].

Example 17:

For refraction at a spherical surface of $n = 1.5, R = 20$ cm, evaluate the aberration for $h = 5$ cm and $h = 7$ cm.

Solution:

From Eq. (4) the computation is direct. For h = 5 cm.

$$SA = \frac{5^2}{2\times 1.5\times 1.5\times 20} = 0.28 \text{ cm}$$

For h = 7 cm, then

$$SA = 0.28 \times \frac{7^2}{5^2} = 0.55 \text{ cm.}$$

Example 18:

An oil immersion objective and a meniscus lens are used in succession, each having $n = 1.60$. Calculate the semi-angles of the incident and emergent cone of rays.

Solution:

For the incident cone, tan θr ÷ (r/n) = n = 1.60 θ = 58°

For the intermediate cone, tan θ' = r ÷ (rn) = 1/n = 0.667 θ' = 33°42'

In the meniscus lens θ' now becomes the incident cone. If θ_1'' is the emergent cone,

$$\tan\theta'' = \frac{1}{n^2}\tan\theta' = 0.261\ \theta'' = 14°36°.$$

Example 19:

Calculate the focal lengths of the crown and flint glass lenses that form an achromatic doublet of focal length 50 cm when placed in contact. The refractive indices for the C and F lines are given below:

Crown	$\mu_C = 1.515$	$\mu_F = 1.523$
Flint	$\mu_C = 1.644$	$\mu_F = 1.664.$

Solution:

$$\omega_1 = \frac{1.523-1.515}{\frac{1.523+1.515}{2}-1} = \frac{0.008}{0.519} = \frac{8}{519}$$

$$\omega_2 = \frac{1.664-1.644}{\frac{1.664+1.644}{2}-1} = \frac{0.020}{0.654} = \frac{10}{327}$$

For achromatism

$$\frac{\omega_1}{f_1}+\frac{\omega_2}{f_2} = 0 \text{ or } \frac{\omega_1}{f_1} = \frac{-\omega_2}{f_2}$$

$$\therefore \quad \frac{1}{f_2} = \frac{\omega_1}{\omega_2 f_1} = \frac{8\times327}{519\times10f_1}$$

$$\frac{1}{F} = \frac{1}{f_1}+\frac{1}{f_2} = \frac{1}{50}$$

$$\therefore \quad \frac{1}{f_1}-\frac{8\times327}{512\times10f} = \frac{1}{50}$$

$$\text{or} \quad \frac{1}{f_1}\left[1-\frac{8\times327}{519\times10}\right] = \frac{1}{50}$$

f_1 = 24.79 cm (crown glass, convex lens)

From equation (1) $f_2 = -\dfrac{519 \times 10 \times 24.79}{8 \times 327}$

$= -$ **49.19 cm.**

Example 20:

Calculate the focal lengths of a convex lens of crown glass (dispersive power 0.012) and a concave lens of flint glass (dispersive power 0.020) so that when placed in contact they form an achromatic converging combination of focal length 30 cm.

Solution:

For achromatism,

$$\frac{\omega_1}{f_1} + \frac{\omega_2}{f_2} = 0$$

Here, $\omega_1 = 0.012$ and $\omega_2 = 0.020$

$$\frac{0.012}{f_1} + \frac{0.020}{f_2} = 0 \qquad \therefore \frac{1}{f_2} = -\frac{0.6}{f_1}$$

Also $\dfrac{1}{F} = \dfrac{1}{f_2} + \dfrac{1}{f_2}$ Bit F = + 30 cm

$$\frac{1}{30\text{cm}} = \frac{1}{f_1} - \frac{0.6}{f_1} \quad f_1 = 30 \text{ cm} \times 0.4 = 12 \text{ cm.}$$

$$\frac{1}{f_2} = -\frac{0.6}{12\text{cm}} \quad f_2 = -20 \text{ cm.}$$

The focal length of the convex lens is 12 cm and the focal length of the concave lens is 20 cm.

Example 18:

Two glasses have dispersive powers in the ratio 2:3. These glasses are to be used in the manufacture of an achromatic objective of focal length 20 cm. What are the focal lengths of the lense

Solution:

For an achromatic objective,

$$\frac{\omega_1}{f_1} + \frac{\omega_2}{f_2} = 0 \qquad \therefore \frac{1}{f_2} = -\frac{\omega_1}{\omega_2 f_1}$$

Also $\frac{1}{F} = \frac{1}{f_1} + \frac{1}{f_2} = \frac{1}{50}$

Here $\frac{\omega_1}{\omega_2} = \frac{\omega_1}{\omega_2}$, and F = 20 cm

$\therefore$ $\frac{1}{f_2} = -\frac{2}{3f_1}$

$$\frac{1}{20} = \frac{1}{f_1} + \frac{1}{f_2}$$

$$= \frac{1}{f_1} - \frac{2}{3f_1} = \frac{1}{3f_1}$$

$f_1 = 203 = $ **6.67 cm.**

EXERCISES

1. Using the theorem in geometry that sum of two sides of a triangle is greater than the third side, give a formal proof of rectilinear propagation of light, based on Fermat's principle.
2. The consider a case (c) in which the mirror falls inside the parabola on one side of M and outside the parabola on the other side. What would you conclude for such a case?
3. State and explain Fermat's principle of stationary time. Derive the laws of refraction using this principle. Give an example where the path of light is a relative maximum rather than a minimum.
4. State and explain Fermat's principle of extremum path and analyze a case where the actual path of light may be a maximum. Use Fermat's principle to deduce Snell's law of refraction.
5. "Fermat's law is a law of extremum path, not only minimum path". Analyze a case where the actual path of light may be a minimum.
6. Parallel to example 4, discuss the case when the refracting surface AO is concave.
7. Consider points P and Q at opposite ends of a diameter of a spherical reflector. Deduce the possible actual paths for rays of light reaching Q from P, stating which path corresponds to a maximum among neighbours and which to a minimum.

8. In example 3 (Fig.2.14) show that the path defined by $\cos^2\theta/2 = 2/3$ is a minimum among neighbouring paths.
9. The dispersive powers of crown and flint glasses are 0.03 and 0.05 respectively. If the difference in the refractive indices of blue and red colours is 0.015 for crown glass and 0.022 for flint glass, calculate the angles of the two prisms for a deviation of 2° without dispersion.
10. Define angular dispersion and dispersive power. Derive the condition to produce dispersion without deviation in a combination of prisms.
11. Can we have dispersion without deviation with a pair of prisms of the same material but of different angles?
12. What do you mean by the terms angular dispersion, angular deviation, and dispersive power of the material? How does dispersive power depend on the refractive index?
13. Derive the condition for the combination of two thin prisms to produce mean deviation without net dispersion. Derive an expression for the net mean deviation.
14. The contact doublets of $f' = 100$ cm using Hard Crown and EDF glasses and achromatic for (i) C and D, (ii) F and H.
15. In Fizeau's toothed wheel could the width of open and opaque spaces be kept unequal? Could we use a rotation speed 10 times that which Fizeau caused?
16. In an experiment with Fizeau's method, using a 500-tooth wheel, the first 'eclipse' occurred at a speed of 1500 revolutions per minute. Calculate :

    (i) The distance between the wheel arid the distant mirror,

    (ii) The speed at which the next 'eclipse' would occur.
17. Comment on the relative merits of Fizeau's and Foucault's methods for measuring the speed of light.
18. A slit source of light, 0.1 mm in width, is placed at the focus of an *ideal* convex lens of $f' = 50$ cm. Deduce the angle between the parallel beams corresponding to the two edges of the slit, and the amount of lateral spread they would show in travelling 20 km distance. Discuss the relevance of this for the intensity consideration in Fizeau's experiment.

19. Describe the Kerr cell method finding the velocity of light in the laboratory. What are the advantages of this method over other methods ?
20. Describe and explain Michelson's rotating mirror method for finding the velocity of light and compare its merits and demerits with those other methods.
21. Light travels to a target and back in 0.5 s, in carbon disulphide. Calculate the distance of the target if the refractive index of the carbon disulphide is 1.46.
22. Describe the Huygens' wave theory?
23. Why light is classified in the category of electromagnetic waves?
24. What is photon theory of light?
25. Describe a modem method for measuring the velocity of light.
26. Describe Anderson's method for determining the velocity of light. What are the merits and demerits of the method?
27. Calculate the time taken by a beam of light to travel through a glass pane of thickness 1 mm. Refractive index of glass is 1.50 and velocity of light through air is 3×10^8 m/s.
28. In measuring the speed of light through a medium the 'comparison' method is preferred over direct measurement. Comment.
29. Explain what the visible region means?
30. Explain the dual nature of light?
31. What is meant by reflection?
32. State and explain the law of reflection.
33. Give a brief account of the methods for finding the velocity of light and give the details of the method which you consider most accurate.
34. Describe in detail Fizeau's method for finding the velocity of light. What are the chief difficulties met with, in carrying out this experiment ?
35. Describe Foucault's method for finding the velocity of light. How does this method justify the correctness of the wave theory of light ?
36. If white light is incident at angle of 30° on a slab of glass that for blue light has a refractive index of 1.7 and for red light of

1.6, what is the angular dispersion between blue and red inside the glass?

37. A single lens magnifier has focal length 5.0 cm for D line, and is used as a × 5 magnifier for that colour. If dispersion from C to F is 0.017, deduce the lateral chromatic aberration

    $| m_C - m_F |$.

38. Dispersive power of a material is sometimes defined as $(n_{violet} - n_{red})/(n_{yellow-1})$. Discuss the need for more precise definition. Explain also why an optician and a photographer require different reference 'colours' for defining dispersive power.

39. The longitudinal chromatic aberration for a convergent lens of crown glass of focal length 1.0 metre, object being considered

    (i) at infinity,

    (ii) at 2 metres in front of the lens.

40. An achromatic contact doublet of focal length 40.0 cm is to be made using hard crown and flint glasses. Calculate the focal lengths of the components from the following data:

    Hard crown $n_D = 1.5190$ $nF - n_C = 0.00852$

    Hint $n_D = 1.5786$ $n_F - n_C = 0.0141$

    Calculate also the geometric factors [$G = (l/R - l/R')$] for the two lenses.